Efficient Algorithms for Infrastructure Networks: Planning Issues and Economic Impact

Frank Phillipson

Reading committee: Prof.dr.ir. K.I. Aardal (Delft University of Technology)
Prof.dr. J.L. van den Berg (University of Twente)
Prof.dr.ir. E.R. van Dam (Tilburg University)
Prof.dr.ir. C.P.M. van Hoesel (Maastricht University)
Prof.dr.ir. J.A. La Poutré (Delft University of Technology)
Prof.dr. A.P. Zwart (VU University Amsterdam)

Copyright © 2014 by Frank Phillipson
Cover photo: TNO project Masterglass - fibre roll out Hardenberg

Typeset by LaTeX.
Printed and bound by Createspace.

ISBN: 978-1494252373

VRIJE UNIVERSITEIT

**Efficient Algorithms for Infrastructure Networks:
Planning Issues and Economic Impact**

ACADEMISCH PROEFSCHRIFT

ter verkrijging van de graad Doctor aan
de Vrije Universiteit Amsterdam,
op gezag van de rector magnificus
prof.dr. F.A. van der Duyn Schouten,
in het openbaar te verdedigen
ten overstaan van de promotiecommissie
van de Faculteit der Exacte Wetenschappen
op donderdag 12 juni 2014 om 15.45 uur
in de aula van de universiteit,
De Boelelaan 1105

door

Frank Phillipson

geboren te Purmerend

promotor: prof.dr. R.D. van der Mei
copromotor: dr. S. Bhulai

Preface

'When do you finally write your PhD-thesis? You have enough material!' This is what Rob van der Mei asked me several times. 'This applies also to Bart [Gijsen] by the way'. And indeed there came a time I was motivated to start writing a thesis and searched for a sponsor. I thought that working on it for one year, one day a week would be enough. Many people said my plan was a mission impossible, but I started in September 2012. It was hard to find the time within my regular working hours. Much of my free time was used for turning the research, which was mostly done already, into papers and material for the thesis. But Rob stayed enthusiastic, believed in me and brought Sandjai Bhulai into the project. So, thank you Rob and Sandjai for bringing me at this point in my academic career.

I like to thank a couple of people who contributed to this work. First my co-authors of the papers that are underlying this thesis: Nadine Croes, Max Schreuder, Timo Leenman, Roland Daamen, Charlotte Smit-Rietveld and Pieter Verhagen. Next my colleagues in projects in which most of the research was done: Antwan Wiegerinck, Carolien van der Vliet-Hameeteman and Karin Spijkers-van Wingerden for the FttCab planning, Stefan Verwijmeren, Jens Cox, Ildiko Theisler and Anja Telkamp for the FttH planning and Harrie van der Vlag, Bas Gerrits and Rob van den Brink for the FttCurb planning. Those projects were funded by KPN, Ministry of Economic Affairs, Grontmij, TNO, Visser & Smit Hanab and VolkerWessels Telecom Infra. I am honoured that Jens and Antwan are my 'paranimfen' for the defence of this thesis.

Doing a project like writing a Ph.D. thesis you cannot do without the support from your family. So thank you, Astrid, Emma, Willemijn, Mum and Dad for all you gave me. I hope that my daughters will get the same opportunities to learn and reach what they want. Remember: *Possunt quia posse videntur*[1] [150].

And Bart, now it is your turn ;-)

Frank Phillipson
Den Haag, April 2014

---

[1]'Ze kunnen het omdat ze denken dat ze het kunnen'. Vrij vertaald staat dat voor: Geloof in jezelf, dan kun je meer dan je denkt.

Contents

Preface i

List of abbreviations v

1 Introduction 1
   1.1 Telecommunication infrastructure 1
   1.2 Electricity infrastructure 3
   1.3 Focus of the thesis 5

2 FttCab planning 7
   2.1 Introduction 7
      2.1.1 DSL-techniques 7
      2.1.2 Migration to FttCab 8
      2.1.3 Literature review 12
   2.2 Sequential approach 14
      2.2.1 Activation 14
      2.2.2 Clustering 22
      2.2.3 Routing 30
   2.3 Integrating clustering and routing 35
      2.3.1 Problem definition 35
      2.3.2 Initial solution 36
      2.3.3 Improvement 42
      2.3.4 Evaluation 43
   2.4 Summary 50

3 FttH planning and economic impact 53
   3.1 Introduction 53
      3.1.1 Hybrid FttH 53
      3.1.2 FttH topologies 54
      3.1.3 Literature review 55
   3.2 Planning of FttCurb 58
      3.2.1 Background 58
      3.2.2 Identifying the options 58
      3.2.3 Standard problems 63
      3.2.4 Cases 66

3.3	FttH planning		70
	3.3.1	Background	70
	3.3.2	Place of the concentration unit	73
	3.3.3	Connecting the houses	81
3.4	Economic impact		92
	3.4.1	Top down description of the techno-economic model	93
	3.4.2	Sub-models in detail	96
	3.4.3	Case 1: migration path	112
	3.4.4	Case 2: total financial impact	117
3.5	Summary		122

4 Electricity network planning — 127

4.1	Introduction		127
	4.1.1	Tactical generation planning	127
	4.1.2	Literature review	130
4.2	Optimal mix of distributed generators		132
	4.2.1	Problem definition	132
	4.2.2	Description of data	145
	4.2.3	Solution method	146
	4.2.4	Results	148
	4.2.5	Implementation	150
4.3	Optimal placing of wind turbines		151
	4.3.1	Power grid modelling	151
	4.3.2	Problem definition	158
	4.3.3	Solution method	163
	4.3.4	Wind power simulation	169
	4.3.5	Results	170
4.4	Summary		177

5 Summary — 181

5.1	Telecom infrastructure	182
5.2	Electricity infrastructure	183

Bibliography — 184

Nederlandse samenvatting — 197

Curriculum Vitae — 203

List of abbreviations

4GBB	Fourth Generation BroadBand
ADSL	Asymmetric Digital Subscriber Line
ARPU	Average Revenue Per User
CFLP	Capacitated Facility Location Problem
CFRS	Cluster First, Route Second
CO	Central Office
CP	Concentration Point
CVRP	Capacitated Vehicle Routing Problem
DG	Distributed Generator
DNPV	Discounted Net Present Value
DSL	Digital Subscriber Line
DSLAM	Digital Subscriber Line Access Multiplexer
EBIT	Earnings Before Interest and Taxes
EDCP	Edge Disjoint Circuits Problem
FLP / UFLP	(Uncapacitated) Facility Location Problem
FttCab	Fibre to the Cabinet
FttCurb	Fibre to the Curb
FttH	Fibre to the Home
GDP	Gross Domestic Product
HV	High Voltage
ISDN	Integrated Services Digital Network
LV	Low Voltage
Micro-CHP	Micro Combined Heat and Power system
MI(Q)P(P)	Mixed Integer (Quadratic) Programming (Problem)
MSTP	Minimum Spanning Tree Problem
MV	Medium Voltage
PON	Passive Optical Network
PoP	Point of Presence
PV	Photo Voltaic
SCFLP	Soft-Capacitated Facility Location Problem
SSCFLP	Single Source Facility Location Problem
TSP	Travelling Salesman Problem
VDSL	Very-high-bit-rate Digital Subscriber Line
VRP	Vehicle Routing Problem

1 Introduction

Electricity and telecommunication network providers operate in a turbulent period. The market is open for competition and the customer demand is changing tremendously, facing the network providers with a huge challenge. In this thesis efficient algorithms are developed for tactical planning of those networks. These algorithms generate fairly accurate solutions within reasonable computation times. In this chapter the background and motivation of the thesis is presented as well as the focus of the research.

1.1 Telecommunication infrastructure

Within Europe, The Netherlands are at the forefront of the broadband penetration of cable internet and ADSL (Asymmetric Digital Subscriber Line) via existing copper networks of cable suppliers and operators. A good broadband penetration is important for a country. In [17] it is estimated that about a quarter of the Gross Domestic Product (GDP) growth in the Netherlands is related to the telecom sector. These effects consist of two components. The first effect, responsible for about five percent, is the *direct* impact of the growth of the telecom capital. The second component is the *indirect* impact of the growth of the telecom capital to GDP growth. The use of such capital by others leads to more innovation and more competition in all markets, including international, leading to 20 percent contribution to the GDP growth in the long term.

To make sure that the networks are capable of delivering the demanded bandwidth and ensuring the position of the Netherlands in broadband penetration, the network operators are expanding their networks. Cable suppliers are busily preparing their networks for the *next generation of broadband internet services*. Telecom operators are also thinking about the future, as the capacity of their copper cables eventually will have its limitations. In order to keep meeting the increasing demand for bandwidth in the future, they also need to bring fibre farther into their networks. In this thesis four topology types are distinguished (see Figure 1.1) and various technologies (like ADSL, VDSL, G.Fast) that can be used within these topologies:

1. Full Copper: services are offered from the Central Office (CO) over a copper cable, using ADSL or VDSL (Very-high-bit-rate Digital Subscriber Line) techniques.

2. Fibre to the Cabinet (FttCab): the fibre connection is extended to the cabinet. From the cabinet the services are offered over the copper cable, using VDSL or G.Fast techniques.

3. Fibre to the Curb (FttCurb or Hybrid FttH): services are offered from a Hybrid FttH Node, which is connected by fibre, close to the customer premises, in the street or in the building. For the remaining copper part G.Fast can be used.

4. Full Fibre to the Home (Full FttH): the fibre connection is brought up to the customer premises.

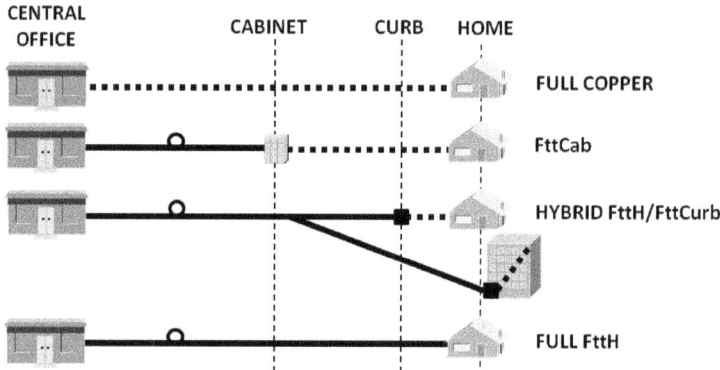

Figure 1.1: Four topologies.

Currently, there are several major fibre projects active within The Netherlands. The fibre roll out started at Onsnet in Nuenen and at the Portaal housing association which connected 50,000 houses to fibre optics [146]. Today, Reggefiber and KPN together provide large parts of The Netherlands with fibre optics. However, the speed of this roll-out is not high enough. The roll out capacity (labour) and the money available for investment in the network are limited. A main part of the premises still only have an ADSL connection with a bandwidth that does not fulfil their demand.

Laying the fibre optic to the cabinet and to each house within a city means that the pavement in a large part of the public streets needs to be opened to install pipes and side-branches to the houses have to be constructed. Next, this new access network needs to be connected to the existing national transportation networks. So when installing this new underground infrastructure, a whole neighbourhood will be under construction. All this digging is very expensive, not only due to the direct investments and construction costs[1] but also indirect costs resulting from the inconvenience[2] caused by it all. In order to limit the inconvenience for residents and companies, it is therefore useful to implement these networks in a smart way. Avoiding social costs will therefore principally result in higher direct costs for implementing the network. To keep the direct costs as low as possible and maintain a good balance between direct and indirect costs, it is necessary to *develop optimization models* and a methodology that will *quantitatively reflect the social costs* as much as possible. This thesis creates a basis for this purpose. With

---

[1]Construction cost are, e.g., the cost for digging, the pipes and cables, equipment.
[2]Inconvenience can be open pavement, detours, loss of turn over.

regard to telecommunication networks there are three main topics that will be studied in this thesis:

1. How to design the network economically?

2. How to manage the building capacity and investment to get a maximum coverage of houses that receive their required bandwidth?

3. How to design the network to minimize social inconvenience?

In Chapter 2 the design of FttCab networks is considered. There exists a lot of related work for fibre network planning. Most of this work takes different network views in terms of the number of network layers and the way of connecting the nodes of the network. None of those existing methods is applicable directly to our case where ring structures are used, the nodes are capacitated, there is a special distance requirement, a minimal penetration rate is used and short calculation times are desired. For this we will introduce new problems and propose proper solution methods.

Next, in Chapter 3 the design of FttCurb and FttH networks, the economic aspects of a migration scheme and the possibilities to minimize social inconvenience when rolling out a FttH network are studied. This work is innovative in relation to other publications in the way that in all those papers G.Fast as technique for FttCurb is not taken into consideration nor are the social inconvenience and the missed revenues due to churn effects.

1.2 Electricity infrastructure

Where telecommunication and cable operators are preparing their networks for dramatic changes in demand, the same applies to electricity network operators. Here a *growing demand*, due to electrical cars, and *high uncertainties*, caused by local production based on natural sources like wind and sun, are expected. This local production is stimulated by the government, as was shown in the energy agreement by market parties and civil organizations in 2013 [128]. They want to reach that at least 1 million households and / or SME companies in 2020 can provide in their own energy demand for a substantial part through sustainable decentralized energy and other forms of renewable generation. Next to this the government wants to achieve that 14% of the energy is produced in a sustainable way in 2020, 16% in 2023 and 100% in 2050. In 2012 this percentage was 4.7% [128].

Local production is merely based on natural sources and reduces the transmission losses caused by central production. Today, every year a staggering 26 billion US dollar is lost due to transportation and distribution of electricity in the US [41]. In a small country like the Netherlands this is still 300 million dollar [33]. Distributed power generation is expected to play a crucial role in reducing transportation loss and meeting future energy demand. In most cases these *distributed generators* (DGs) are based on renewable energy such as Photo Voltaic (PV) solar cells and wind turbines, but there are also new technologies to increase efficiency such as the micro Combined Heat and Power (micro-CHP) systems. Another reason why distributed generation, especially from renewal sources, is becoming more attractive is that it can provide security of energy supply and contribute to a diverse energy portfolio. Concerns regarding environmental issues, in particular global warming and exhaustion of non-renewable energy resources, are also on the rise.

Although the introduction of DGs looks promising, the challenge will be to integrate the increasing number of (small) generation units in an electricity system that up to now has been very centralized, integrated and planned (see, e.g., [115]). Since most DGs rely on exploitation of natural sources of energy they exhibit high fluctuations in production over time. This means that electricity generated by DGs will probably not match load demand and can cause over- or underproduction of electricity. This uncertainty is undesirable in a situation where we are highly dependent on energy and more and more dependent on electronic devices.

The current technological solution to solve possible transport problems is to reinforce the existing grid as stated by Gitizahed [51]. This is very expensive and is avoided as much as possible. Another solution, still in development, is to make the grid smarter by controlling fluctuations in production and consumption, using smart grids (see Kok [80, 81]) and production planning (see for example Kopanos et al. [84]). Next to this, many trials have been done by managing the behaviour of customers, like in Faruqui et al. [45].

Before grid reinforcements are realized and smart grids that can perform sophisticated load balancing are created, network operators should have some idea about the global demand and supply characteristics and they should perform *tactical planning*: try to find an optimal mix and placement of DGs while keeping network capacities in mind. They have to make sure that all demands are satisfied, by power plants and/or DGs. Each DG type has a different (stochastic) production pattern depending on, e.g., the sun, wind or heat demand, that may complement each other. When there is sunshine (or wind) all PV solar panels (or micro wind turbines) in the district will generate electricity at the same time, leading to a sudden increase in generated electricity.

The two important problems the electricity operators are facing are thus a change in demand and high uncertainties. They need to enhance their network capacity and come up with strategies to balance the supply and demand on their networks. This balancing has two aspects: (1) Those generators based on natural resources are highly volatile, and (2) Their profile do not necessarily match the demand profile. These problems are studied from a different perspective:

1. What combination of local generators should be placed in a district to match the demand closely, such that the total energy loss, mainly due to transportation and storage, is minimized?

2. Can local generators, for instance wind mills, be spread over the country such that their joint volatility is less but also the transportation losses are minimized?

These problems will be the topics of Chapter 4 of this thesis. To the best of our knowledge we are the first developing a model for tactical DG planning, for size and nature, within a district, minimizing total transportation loss and avoiding overload, the first problem above. Kools [83] mentions three important issues when evaluating the effect of DGs: (1)

the capacity of the generators, (2) the type of the generators and (3) the location of the generators. These issues are dealt with in various combinations in literature. The first problem focusses on the combination (1) and (2) what has not been done so far. The second problem focusses on (1) and (3), but introduces the uncertainty and correlation of wind, which is in our view innovative.

1.3 Focus of the thesis

In practice planners of infrastructure networks are primarily interested in interactive planning tools, for which the computation times are short (seconds, minutes at most). There are several reasons for this attitude:

- The real life situation is far more complex than can be modelled in the planning tool. After the planning has been created by the tool, the geographical conditions need severe site surveys that influence the final planning significantly and the geographical information planners have is often unreliable. Typical questions here are: 'Can we place the drilling rig in that street?', 'Do we get a permit for that street?' and 'Is that cable really existing?'. These modifications to the 'optimal' planning have a major impact on the final planning that void the performance gain of some optimal solution completely.

- Methods based on 'rules of thumb' are better to understand and are closer to the way the planner work, what helps the credibility of the tool. And perhaps most importantly: a tool that is not trusted by the user, is not used at all.

- Large companies have strict IT policies that allow to use only standard applications. Sometimes the only way to run a tool is using Excel and VBA (Visual Basic for Applications), what give restrictions on the used methods in the implementation.

- The planning is not made just once. Most tools are tactically used for tactical scenario research and the search for parameter settings. Multiple runs are needed for this. In a more operational environment the planning for an area is done sometimes caused by adding (detailed) information and there are hundreds or thousands of areas to plan.

The above reasons raise the need for planning algorithms that are *efficient, scalable, easy to implement* yet *fairly accurate*. The scalability is important, where the size of the areas can vary from a little village to a crowded (part) of a big city. The idea of an interactive planning tool and thus the restriction to the calculation time, will be used in the remainder of the thesis. These complex planning problems will be studied and solution methods are presented that were created and, in most cases, implemented for practical use by network planners.

2   FttCab planning

More digital services come available to customers every day and even more importantly, these services are asking more bandwidth. This is mainly due to the integration of video, high definition and 3D, into numerous services that are used simultaneously. Bandwidth demand grows approximately 30% to 40% annually between now and 2020 on fixed connections [146]. Telecom operators have to make their access networks ready for this. Therefore they have to make the costly step to FttCab, where the services are offered using the VDSL technique from the cabinet, or, even more costly, the step to FttH. The roll out of FttH will be taking too long to compete with the cable TV operators, who can offer the required bandwidth at this moment. For many telecom operators bringing FttCab in the next years will be the only way to survive. In this chapter an efficient three step approach for FttCap planning is presented. This chapter is based on [36, 117, 121, 122].

2.1   Introduction

In this section some technical background on DSL (Digital Subscriber Line) techniques is provided, the problem is sketched and placed in the existing literature.

2.1.1   DSL-techniques

The twisted pair copper network was implemented for analogue phone service, and consists of large numbers of copper wire pairs which originated bundled in the Central Office (CO). This copper network facilitated telephone, telefax and facsimile services until well into the 1980's, as well as narrow banded data communication between computers by means of analogue speech modems. The digitalization of the exchanges was well under way during the 1980's, and at the end of that decade this became noticeable for subscribers with the introduction of ISDN (Integrated Services Digital Network). As of 1999, DSL development was introduced to the consumer market with ADSL. ADSL and the following generations with ever increasing data speed (ADSL2, ADSL2+) could be offered to the consumer at affordable rates, thereby giving the development of broadband in The Netherlands a huge boost as well.

Thick copper wires leave from a CO to the end users. In The Netherlands, these distribution cables usually consist of 900 copper wire pairs, which continuously branch out until a cable pair is delivered to the end user. Usually there is a cable divider located in a cabinet between the CO and the premises, and sometimes even two cabinets are between the CO and the premises (cascaded COs). Sometimes a cabinet is skipped for very short distances and the end user is connected to the CO directly. Every end user has a minimum of one personal copper wire pair at his disposal, but incidentally there are even multiple wire pairs and/or spare wire pairs which may be made available by a technician. Such a wire pair is also called a *twisted pair*, as it concerns two intertwined copper wires.

So-called DSL Access Multiplexers (DSLAMs) from a DSL operator are set up in a CO, which divides the traffic from the core network over the connected DSL lines (multiplex) and vice versa. Every line originates at the DSLAM and terminates at the end user's DSL modem. With ADSL, the complete line runs over copper. By now DSL technology consists of the following types and generations (in particular those which are relevant for the consumer market):

- ADSL: Asymmetric Digital Subscriber Line.

- ADSL2: Successor of ADSL, with improvements.

- ADSL2+: Successor of ADSL2, with higher capacity.

- VDSL: Very high bit rate speed DSL: new generation DSL technology optimized for implementation from the cabinet.

- VDSL2: Successor of VDSL with numerous improvements to be also used next to ADSL2+.

The typical maximum speed, depending on the copper length of the connection is depicted in Figure 2.1. In practice this relation is not that smooth: all kinds of other factors will influence the speed the user experiences. Think of the quality of the cable and crosstalk effects. The relatively new technique of *vectoring* will provide a more smooth line, where this technique aims on reduction of crosstalk levels and improvement of performance. It is based on the concept of noise cancellation. Another method to increase the user's bandwidth is *bonding*, where the user gets two DSL connections, using the, in most cases already existing, second wire pair.

An important aspect of the subsequent DSL technology generations is that an ever increasing capacity can only be realized over ever decreasing lengths of the copper wire. A result of this effect is that the fibre optic needs to be located closer to residences in order to offer every household the opportunity to benefit from a higher capacity from a technology like VDSL2. The final step in this migration process is the one in which the fibre optic is installed to the border of the property and only the last bit of copper wire (over the property or inside the residence) is used. With that currently missing step it is almost a complete FttH solution. So in principle, a step by step migration to a final FttH stage may be realized with DSL technology.

2.1.2 Migration to FttCab

When migrating from an ADSL based network to a VDSL based network, a choice can be made between two options. The first option is to offer *VDSL from the CO*. This requires a relatively small investment, as the location is already connected to a fibre

2.1 Introduction

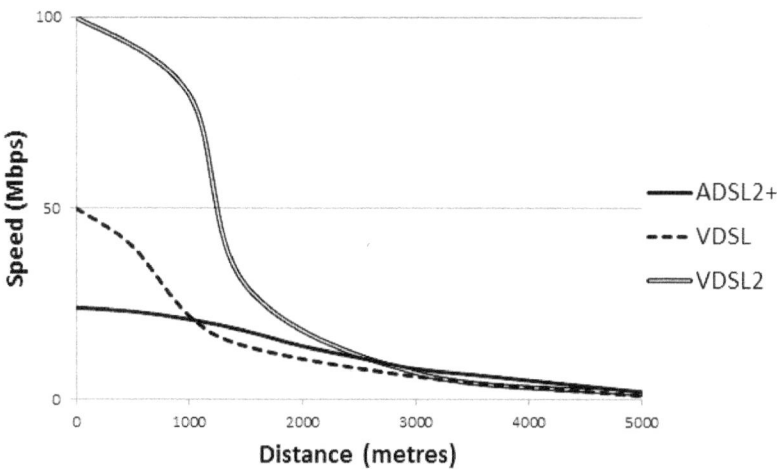

Figure 2.1: Typical DSL speed as a function of the distance.

optic network; only the end-user modems need adjustment. This is a viable option for residences which are situated less than 1 km copper distance from the CO. This 1 km is the cut-off for the added value of VDSL (see Figure 2.1). For residences situated farther from the CO, it will be necessary to extend the fibre optic farther into the direction of the houses. This is the second option, in which the cabinet is selected as the next logical active point leading to *FttCab*. This step has to be made by Telecom operators to meet the continuous growth of bandwidth demand of their customers. An alternative will be the roll out of FttH, but that will take too long to compete with the cable TV operators, who can offer the required bandwidth to the majority of their customers at this moment. For many telecom operators bringing VDSL, partly using FttCab, in the next two years will be the only way to survive. However, FttCab can even be an useful intermediate step towards FttH as is shown in Section 3.4.

Connecting the cabinet with a fibre optic cable and installing the necessary hardware into it will be referred to as *activating a cabinet* from this point onwards. When looking at FttCab, an operator does not want to activate all cabinets but only a selection. The operator wants to reach as many customers with as little investment as possible; usually the choice is made for a minimal penetration of, for example, 85%. The operator will therefore look for a minimal cost selection of activated cabinets that collectively have more than 85% of the customers within 1 km. The cabinets that are not activated will be connected to an activated cabinet (partly) using existing copper connections and are called *placed in cascade*. Still, customers connected to these cabinets can be within 1 km from the activated cabinet and hence use VDSL. The activated cabinets have to be connected to the CO via new fibre optic cables. This can be done by a star, tree or ring structure. Due to reliability constraints in the Netherlands the choice is made for rings, and even broader as Tipper et al. [145, 148] and Gong et al. [57] sketch the need for redundancy in the network, even in access networks. The next question then is which cabinet is served by which fibre optic ring. Each ring that is constructed should meet a

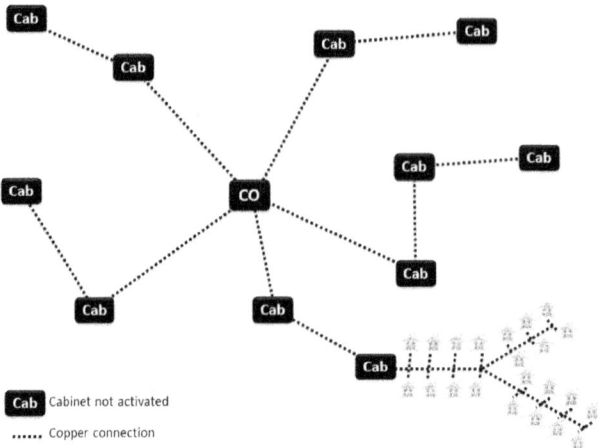

Figure 2.2: Starting point of the planning.

capacity constraint and should be edge disjoint to guarantee the desired reliability. In other words, it is not allowed that a trench or pipe is used twice within one ring.

To summarize, the operators typically want a solution that meets the following requirements:

1. A ring-shaped network structure is used to guarantee a high reliability.

2. Each ring has a maximum capacity, here a number of customers that can be included in one circuit.

3. Each cabinet will be included in only one cluster. This is a logical property of the network design, since it would just be too expensive to include every cabinet in multiple circuits.

4. Each ring that is constructed should use every track only once, in mathematical term, it has to be *edge disjoint*. In other words, it is not allowed that an edge is used twice within one ring. Note further that between rings, no edge disjointness is required. Multiple rings can, therefore, make use of the same pipe. Another remark to be made is that a ring is not required to be node disjoint.

If the operator wants to migrate to FttCab, he has to design and plan the new network, starting with the already available equipment and cables from the existing network. In this chapter first a three step approach is presented for this planning problem. These three sub-problems in our approach are:

1. Which of the cabinets should be activated in order to reach the desired percentage of households at minimal costs? Figure 2.2 shows the starting point for a small example. All cabinets (Cab) are connected through copper to the CO. Several residences are connected to the cabinet; this is only shown for one cabinet in the

2.1 Introduction

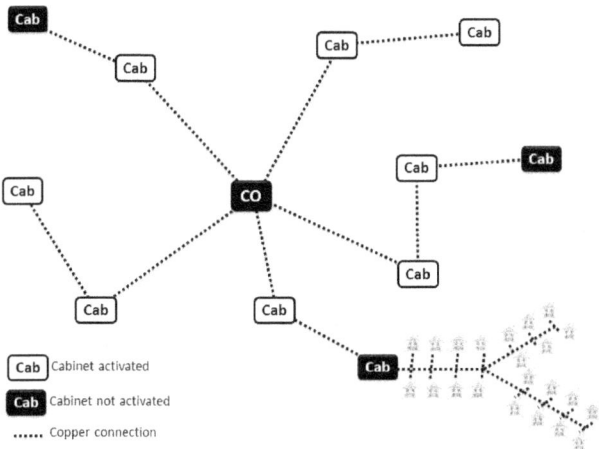

Figure 2.3: Which cabinets must be activated?

illustration. Now a subset of the cabinets needs to be activated in order to reach the intended number of households over copper from an activated cabinet within the set distance (see Figure 2.3).

2. Which cabinet is served by which fibre ring? The cabinets now have to be partitioned into groups in order to determine which cabinets will be jointly connected by one fibre ring (see Figure 2.4).

3. How will each fibre ring run? The physical route of the fibre rings needs to determined. What order will they be connected in, and how does this route run taking into account the fact that one ring cannot use the same fibre or duct twice and the possibility to reuse existing infrastructure? See Figure 2.5.

The approach presented in this chapter of the thesis was used in a tactical planning tool of a telecom operator. Recall that in practice planners are primarily interested in an interactive tool that generates fairly accurate solutions. This raised the need for an interactive planning tool and thus the restriction to the calculation time. This consideration is the main reason for the partitioning of the problem in the above mentioned three parts. Three difficult problems are recognized here: a generalisation of the Facility Location Problem (the first part), some k-Means Clustering Problem (second part) and a Travelling Salesman Problem (third part). The last two together can be seen as a Vehicle Route Planning (VRP) problem with an extra constraint due to the edge disjointness within the rings. In fact here a Cluster First-Route Second method is chosen for this problem. Other known approaches like the savings algorithm of Clarke and Wright [28] and Route First-Cluster Second are less useful here due to the extra (edge disjointness) constraint. Other approaches, like meta-heuristics, are expected to result in much larger calculation times. The overview of Laporte et al. [90] gives an indication for that. A greedy insertion method would be the only viable alternative.

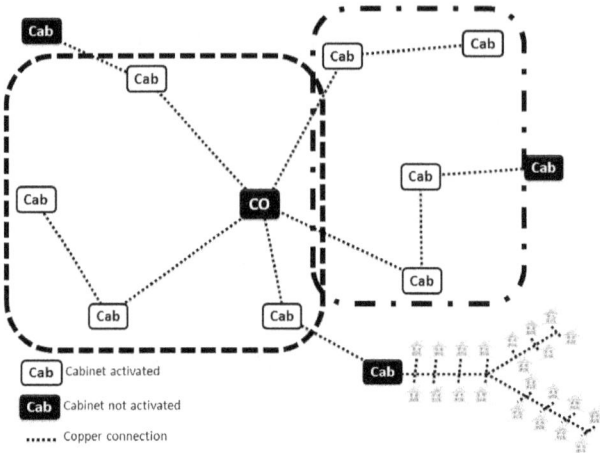

Figure 2.4: Which cabinets are together on one ring?

In the remainder of this chapter this three step approach is presented in Section 2.2. Next, the combined approach of the two last steps, using a greedy insertion method is defined and the results of that approach are compared to the results of the separate steps, in Section 2.3.

2.1.3 Literature review

With respect to network planning there is a lot of related work. Most of this work takes different network views in terms of the number of layers and the way of connecting. For the latter there are design problems consisting of determining graphs that represent network topology for stars, double stars and doubly connected trees, as also stated by Chardy et al. [25]. Here a short overview of other publications regarding network planning is given, but none of the methods is applicable directly to our case where ring structures are used, the nodes are capacitated, there is a special distance requirement, a minimal penetration rate is used and short calculation times are desired.

Kalsch et al. [75] consider almost the same problem, also combining clustering and routing of nodes. They developed a mathematical model and a heuristic, based on cluster first - route second, to create ring structures in a network. Their solution could be used for our problem, however it is quite unclear from [75] how the first part (clustering) is solved and the computation time seems too long for our approach. Henningsson et al. [62] also tried to combine routing and clustering. They chose a Lagrangian Based Column Generation approach, solving problems up to 70 nodes in calculation times that vary from 21 to 8031 seconds. Too long for our interactive approach.

Gollowitzer et al. [55] present the Two Level Network Design (TLND) problem for greenfield deployments and roll-out mixed strategies of FttH and FttCurb, i.e., some customers are served by copper cables, some by fiber optic lines. For two types of customers (primary and secondary), an additional set of Steiner nodes and fixed costs for

2.1 Introduction

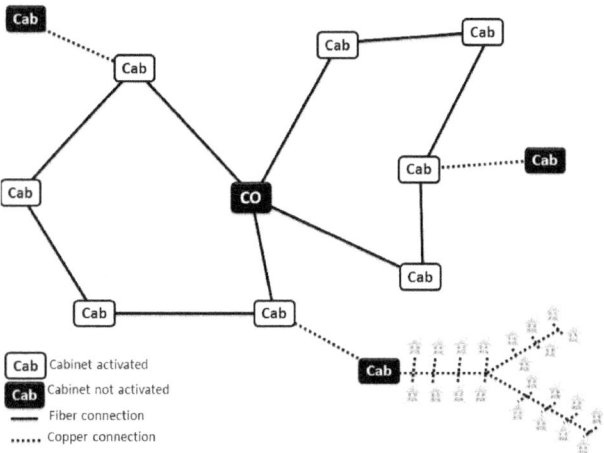

Figure 2.5: Physical route of the fibre ring.

installing either a primary or a secondary technology on each edge, the TLND problem seeks a minimum cost connected sub-graph obeying a tree-tree topology, i.e., the primary nodes are connected by a rooted primary tree; the secondary nodes can be connected using both primary and secondary technology. In [55] an important extension of TLND is presented in which additional transition costs need to be paid for intermediate facilities placed at the transition nodes, i.e., nodes where the change of technology takes place.

Gollowitzer and Ljubić [54] give an overview of Mixed Integer Programming (MIP) models for connected facility location problems. Here also only tree structures and uncapacitated nodes are considered.

Mateus et al. [105] describe the network design problem of locating a set of concentrators which serves a set of customers with known demands. The uncapacitated facility location model is applied to locate the concentrators. Then, for each concentrator they analyse a topological optimization of its sub-network based on a simple heuristic. In a third phase, they solve the upper level sub-network connecting the concentrators to a root node in a tree structure.

Mitcsenkov et al. [109] address broadband PON (Passive Optical Network) access network design minimizing deployment costs using a heuristic solution. The questions here are: how to form groups of customers that share a PON splitter, where is the splitter placed, what is the best path from the customer to its splitter unit and how to connect splitters to the central office. The first problem is regarded as a clustering problem which they solve heuristically by combining nearby shortest paths from the customer to the CO. The second problem is solved optimally by just calculating the optimal location from the (small) set of possible locations. The last two problems are solved by a Steiner Tree problem, using a 2-approximation heuristic, the Distance Network Heuristic as presented by Kou and Berman [86].

In [155] the topology design of hierarchical hybrid fibre-VDSL access networks is presented by Zhao et al. as a NP-hard problem. A complete strategy is proposed to find a cost-effective and highly-reliable network with heuristic algorithms in a short time (up to 700 seconds). The Ant Colony Optimization has been implemented for a clustering problem. The network structure they look at is a two layer cabinet solution with Branch Micro Switches (BMS) and Lead Micro Switches (LMS) where the BMS is connected with the CO with two paths and the LMS is connected to two BMSs. The users are connected in a star with one LMS. The major planning problem here is to build up the intermediate BMS level.

Gódor and Magyar [52] look at the access planning problem from the side of mobile network planning; here the facility location problem is seen as the basis for solving this. A two-phase heuristic planning algorithm is used. In the first phase, they construct an initial network-solution with a K-means algorithm and in the second phase, they improve it by moves or swaps.

Gendreau [50] and Fink [47] present heuristics for the (General) Ring Network Design Problem. They deal with the construction of one single ring. Revenues are defined for each pair of nodes included in the ring and construction costs for each direct link in the ring are used. The objective is to maximize the sum of all revenues minus the construction cost when building the ring. However, they do not look at multiple rings and do not give a solution for the complicating issues around pre-existing pipes and edge disjointness.

Thomadsen and Stidsen [143] present the hierarchical ring network design problems. This problem consists of clustering, hub selection, ring design and routing problems, in many ways comparable to our problem. In their paper a branch-and-price algorithm is presented for jointly solving this underlying problems. However, also their calculation time is much higher than our goal and the size of their example much smaller.

Baldacci et al. [10] introduced the Capacitated m-Ring-Star Problem (CmRSP) based on the Ring Star Problem which was already introduced by Labbé [88]. They consider a ring-star topology, where the rings ar node-disjoint and Steiner nodes can be used to create the link between the ring and a branch of the tree. The CmRSP does not include however the use of existing pipes into the network.

2.2 Sequential approach

In this section the three step approach as presented in the previous section is studied. In each sub-section one of the steps is presented: activation, clustering and routing. Each sub-section consists of a formal formulation of the problem, some known solutions for this problem in literature, our proposed approach and an evaluation of the performance on some real life cases.

2.2.1 Activation

The *cabinet activation problem* is the question which cabinets should be provided with fibre and active equipment if the operator wants to reach as many customers with as little investment as possible. A household is reached when the distance over copper is less than a chosen length, typically one kilometre. Households which meet this requirement

2.2 Sequential approach

are said to meet the distance requirement. To this end, the problem is formulated as an Integer Programming Problem (IPP). Since the problem is NP-hard, a heuristic to approximate the optimal solution is developed. Extensive experiments with a number of real-life data sets of Dutch cities have been performed to benchmark the results against the optimal solution. The results show that the heuristic performs extremely well, and leads to close-to-optimal solutions with negligible computation times.

2.2.1.1 Problem definition

Here the problem is described as a mathematical model, for this the Partly Covered Single Source Capacitated Facility Location Problem (pcSSCFLP) is defined. The chosen structure is that of an Integer Programming Problem. The decision variables for this problem are (for $i, j = 1, \ldots, n$):

$$x_{ij} = \begin{cases} 1 & \text{if cabinet } i \text{ is connected to cabinet } j, \\ 0 & \text{otherwise.} \end{cases}$$

$$y_j = \begin{cases} 1 & \text{if cabinet } j \text{ is activated}, \\ 0 & \text{otherwise.} \end{cases}$$

The model input is described by the following parameters:

b_{ij} = number of clients within a chosen length if cabinet i is connected via cabinet j, b_{ii} is the number of customers at location i, namely, if this location is activated itself, the loss of customers is 0.

c_{ij} = connection costs (in some currency) if cabinet i is connected via cabinet j, c_{ii} is the activation costs of cabinet i, the costs to place a cabinet with the equipment inside.

D = number of clients that has to be within the chosen length.

w_i = maximum number of clients on cabinet i.

t_i = maximum number of cascades on cabinet i.

The optimisation problem is formulated as follows:

$$\min \sum_{i=1}^{n} \sum_{j=1}^{n} c_{ij} x_{ij} + \sum_{j=1}^{n} c_{jj} y_j, \tag{2.1}$$

under the following constraints:

$$\sum_{j=1}^{n} x_{ij} = 1 \quad (i = 1, \ldots, n), \tag{2.2}$$

$$\sum_{i=1}^{n} \sum_{j=1}^{n} b_{ij} x_{ij} \geq D, \tag{2.3}$$

$$\sum_{i=1}^{n} x_{ij} \leq t_j \quad (j = 1, \ldots, n), \tag{2.4}$$

$$\sum_{i=1}^{n} b_{ii} x_{ij} \leq w_j y_j \quad (j = 1, \ldots, n), \tag{2.5}$$

$$x_{ij}, y_j \in \{0, 1\} \quad (i, j = 1, \ldots, n). \tag{2.6}$$

The constraints are explained by:

(2.2) Each cabinet has to be dealt with via exactly one (other) cabinet.
(2.3) The total number of customers connected within the chosen length, for example 1 kilometre, to the cabinet needs to be larger than or equal to D.
(2.4) No more than t_j cabinets may be cascaded to one other cabinet.
(2.5) No more than w_j customers may be within a cascade if cabinet j is activated.
(2.6) Both $x_{i,j}$ and y_j are binary variables.

Theorem 2.2.1. *The pcSSCFLP is NP-hard.*

Proof. Setting $t_i = \infty$, $i = 1, \ldots, n$, and $D = \sum_{i=1}^{n} b_{ii}$ results in the Single Source Capacitated Facility Location Problem as defined and proven NP-hard in Darby-Dowman and Lewis [37]. □

2.2.1.2 Solution

As the problem is NP-hard a heuristic approach is proposed. The idea behind this heuristic approach is that in a greedy way is searched for the cascades that deliver the highest savings and next these cascades are realized until adding an extra cascade violates the condition that a certain number of customers must be connected within the chosen length. Next the solution of this greedy heuristic is improved by performing a 2-opt step: search for two cabinets and swap their position such that the total solution gets better. The proposed heuristic is shown schematically in Figure 2.6 and in detail in Algorithm 1. Here the heuristic is described with reference to the steps in Figure 2.6 and the lines in Algorithm 1:

1. (Input) Start with the solution in which all cabinets are activated. In the case this solution does not meet the desired total number of customers connected within the chosen length, the algorithm stops.

2. (Line 1-5) All possible cascade arrangements are determined and the savings of this arrangement as well as the number of customers which as a result are positioned outside the desired distance of, for example, 1000 meter are reviewed. Note that $i = j$ is possible in all steps. If the cascade (i, j) is constructed, the activation costs of i is saved, but the costs of connection (i, j) are incurred, resulting in $(\delta cost)_{ij}$ in the first equation. Next, some customers might fall out of the required distance, described by $(\delta cust)_{ij}$. This will be the number of customers connected to i, minus the number reached if the connection is made. The last equation, $cust_{ij}$ counts the number of total customers if this (and only this) cascade is realized.

2.2 Sequential approach

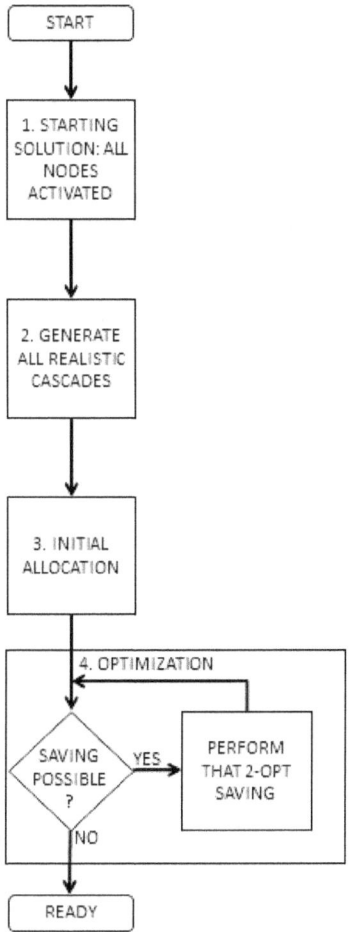

Figure 2.6: Overview of the heuristic.

3. (Line 6) The solutions that generate a saving are sorted on the value $\delta cost_{ij}$.

4. (Line 7-13) Realize the solutions with the largest saving, until the requirement of e.g. 90% of the customers is reached. Here the constraint in Line 9 make sure that:

   - Cabinet i has not been assigned to another cabinet yet.
   - Cabinet j has still room for one more cascade.
   - Enough customers are still within the distance requirement.
   - The capacity of cabinet j is not violated.

   If all these requirements are met, the cascade is realized.

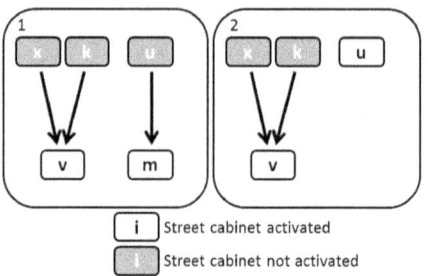

Figure 2.7: Possible situations before swap.

5. (Line 14-37) Perform a 2-opt approach to improve the solution. Here is searched for an allowed cascade: find the first candidate for the swap that was not realized in step 3, but would lead to cost savings. This means that an existing cascade has to be broken and one have to find the best allowed (second) candidate that makes room for the intended cascade and gives an overall cost savings. If this second candidate is found, the swap is realized. If this results in a swap, a new improvement has to be found; if no swap can be found the best solution is found and the heuristic ends. Figure 2.7 gives two possible starting positions: cabinets v and m are activated. Both x and k are connected to v. Cabinet u is connected to m, or u is activated itself. Here is assumed that v had its maximum cascades ($t_v = 2$). If one of the situations in Figure 2.8 can be found that is better, then that swap has to be performed. The set K in Line 23 contains all the cabinets that now are in cascade with v and as such a candidate for the swap, where we want to connect u to v. In Figure 2.7 these are x and k. The set L contains all possible new cascades for the cabinets in K. In Figure 2.8 this would be y and m.[1]

2.2.1.3 Evaluation

The heuristic was tested on five real life cases in the Netherlands. This section presents the approach and results of these tests.

2.2.1.3.1 Approach

To test the presented heuristic five real life cases of different size of number of cabinets were created. The cases involve all Dutch cities and are presented in Table 2.1. The cities vary in the number of inhabitants, the number of clients per cabinet, the number of cabinets and the nature of the city (rural or highly urbanized). The cases were solved in three ways:

1. with AIMMS[2]/CPLEX[3],

---

[1] Activating k can not be a cheaper alternative, where as step 4 would have chosen this before if that was true.

[2] AIMMS stands for 'Advanced Interactive Multidimensional Modeling System' and is a software system designed for modeling and solving large-scale optimization and scheduling-type problems.

[3] CPLEX is a solver, used by AIMMS to solve the optimization problems.

2.2 Sequential approach

Algorithm 1 Activation

Require: A starting solution: the situation that all cabinets are activated:

$$x_{ij} = \begin{cases} 1 & \text{if } i = j, \\ 0 & \text{else.} \end{cases} \quad y_j = 1 \quad (i, j = 1, \ldots, n).$$

Ensure: A solution (x, y) where the total activation costs are minimized.
1: **for** all combinations of (i, j) **do**
2: $\quad (\delta cost)_{ij} = c_{ij} - c_{ii}$;
3: $\quad (\delta cust)_{ij} = b_{ii} - b_{ij}$;
4: $\quad cust_{ij} = b_{ii} + b_{jj}$;
5: **end for**
6: make a sorted list $p = (p_{(1)}, p_{(2)}, \ldots)$ with item $p_{(1)} = \{(i,j) | (\delta cost)_{ij} \leq (\delta cost)_{kl} \forall (k,l); \delta cost_{ij} \leq 0; cust_{ij} \leq w_j\}$ the feasible solution with the highest saving, $p_{(2)}$ the next highest saving and so on.
7: **for** $t = 1, \ldots, |p|$ **do**
8: $\quad (i, j) = p_{(t)}$;
9: $\quad$ **if** $y_i = 1$ and $\sum_i x_{ij} < t_j$ and $\sum_{ij} b_{ij}x_{ij} - D + (\delta k)_{ij} \geq 0$ and $\sum_k b_{kk}x_{kj} + b_{ii} \leq w_j$ **then**
10: $\quad\quad y_i = 0$;
11: $\quad\quad x_{ij} = 1$;
12: $\quad$ **end if**
13: **end for**
14: **for** $t = 1, \ldots, |p|$ **do**
15: $\quad (u, v) = p_{(t)}$;
16: $\quad m = \sum_k k * x_{uk}$;
17: $\quad (\delta swapcost)_{umv} = c_{uv} - c_{um}$;
18: $\quad (\delta swapcust)_{umv} = b_{uv} - b_{um}$;
19: $\quad$ **if** $x_{uv} = 0$ and $y_v = 1$ and $\delta swapcost)_{umv} < 0$ **then**
20: $\quad\quad$ break for loop;
21: $\quad$ **end if**
22: **end for**
23: $K = \{k | x_{kv} = 1\}$; $L = \{l | (k,l) \in p\}$;
24: **for** all combinations (k, l) where $k \in K$ and $l \in L$ **do**
25: $\quad (\delta swapcost)_{kvl} = c_{kl} - c_{kv}$;
26: $\quad (\delta swapcust)_{kvl} = b_{kl} - b_{kv}$;
27: **end for**
28: **if** $y_l = 1$ and $\sum_i x_{il} < t_l$ and $\sum_{ij} b_{ij}x_{ij} - D + (\delta swapcust)_{kvl} + (\delta swapcust)_{umv} \geq 0$ and $\sum_i b_{ii}x_{il} + b_{kk} \leq w_l$ and $\sum_i b_{ii}x_{iv} + b_{uu} - b_{kk} \leq w_v$ and $(\delta swapcost)_{kvl} + (\delta swapcost)_{umv} < 0$ **then**
29: $\quad swapcost_{kluv} = (\delta swapcost)_{kvl} + (\delta swapcost)_{umv}$;
30: **else**
31: $\quad swapcost_{kluv} = \infty$;
32: **end if**
33: $(k, l, u, v) = \text{argmin } swapcost_{kluv}$;
34: $y_u = 0$; $x_{um} = 0$; $x_{uv} = 1$; $x_{kv} = 0$; $x_{kl} = 1$;
35: **if** a swap has been performed **then**
36: $\quad$ go to line 14
37: **end if**

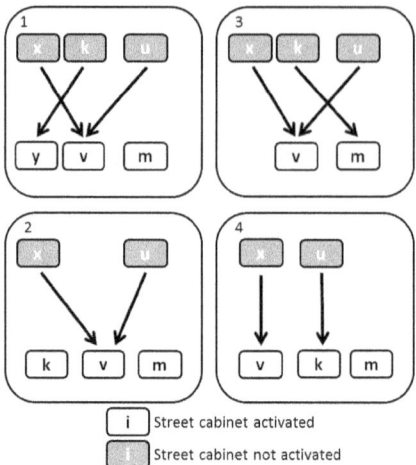

Figure 2.8: Possible situations after swap.

2. with our proposed heuristic approach implemented in GIANT-PlanXS[4],

3. with a random insertion method,

all on a Intel Core i5-2520M Processor with 4 GB RAM. The random insertion was taken as a comparison how a simple method would perform.

The random insertion method works as follows:

1. Determine all possible cascades: all clients within 1 kilometre of the activated cabinet and the cascade gives a cost saving.

2. Effectuate, at random, one of these cascades. Say cabinet i is activated and cabinet j is placed in cascade with i.

3. Update the list of possible cascades:

   - All cascades where cabinet i is placed in cascade are removed.
   - All cascades where cabinet j is placed in cascade are removed.
   - All cascades where cabinet j is activated are removed.
   - All cascades where cabinet i is activated are removed if cabinet i reached its maximum number of cascades (t_i).

4. If there are possible cascades go to step 2; else end.

This was repeated 1000 times for each instance.

---

[4]Giant-PlanXS is a TNO tool, based on Matlab.

2.2 Sequential approach

City	Cabinets	Subscribers	Inhabitants	Number of COs
Venray	54	13,646	43,000	1
Delft	130	32,466	99,000	1
Amstelveen	180	38,868	85,000	1
Amsterdam	528	70,884	790,000	1
Den Haag	866	271,622	502,000	6

Table 2.1: The selected cases for testing.

	AIMSS		Heuristic		Random	
City	Value	Time	Value	Time	Value	Time
Venray	911538	50 s	918344	< 1 s	959265 (1010502)	< 1 s
Delft	1665492	276 s	1698157	< 1 s	2115270 (2348213)	18 s
Amstelveen	2377931	245 s	2461688	< 1 s	3089370 (3257735)	45 s
Amsterdam	4988354	1745 s*	5586957	< 1 s	9619660 (9933367)	1376 s
Den Haag	10813254	1642 s	11070336	< 1 s	16339446 (16843820)	3980 s

Table 2.2: Results for $t_j = 4, j = 1, \ldots, n$.

2.2.1.3.2 Numerical results

Table 2.2 represents the results for the cases where the maximum number of cabinets in the cascade equals four ($t_j = 4$ for $j = 1, \ldots, n$) for all cabinets. For all methods the best value of the objective function is presented and the time to find the solution. The random insertion method presents the best value found in 1000 replications and shows the average score, of the 1000 replications, between brackets. For example, the case 'Venray' is solved by AIMMS in 50 seconds (both building the model and solving it) and by both the heuristic and the random insertion method in less than 1 second. Table 2.3 shows the results for a maximum number of 2 cabinets ($t_j = 2$ for $j = 1, \ldots, n$) in the cascade. The solution of the city of Delft is shown in Figure 2.9, where the cascades are shown with thick lines.

The calculations where the time is marked with a (*) are not completed due to memory problems in AIMMS. The solution here is not necessarily optimal. The heuristic approach delivers a result very quickly for all cases, in all cases within 1 second, whereas CPLEX runs for minutes for most cases and the random insertion method grows quickly with the number of cabinets. The found solution of the heuristic is within reasonable distance of the optimum, see Table 2.4. Here the solution relative to the case that

	AIMSS		Heuristic		Random	
City	Value	Time	Value	Time	Value	Time
Venray	913232	32 s	926310	< 1 s	961121 (1009696)	< 1 s
Delft	1735156	261 s	1758314	< 1 s	2143430 (2349740)	17 s
Amstelveen	2425093	102 s	2472163	< 1 s	3047640 (3260668)	45 s
Amsterdam	5506922	4248 s*	5824534	< 1 s	9573190 (9931352)	1373 s
Den Haag	11366196	3402 s*	11500549	< 1 s	16195961 (16837942)	3971 s

Table 2.3: Results for $t_j = 2, j = 1, \ldots, n$.

	$t_j = 4, j = 1, \ldots, n$			$t_j = 2, j = 1, \ldots, n$		
City	AIMMS	Heuristic	Random	AIMMS	Heuristic	Random
Venray	81.5%	82.1%	85.8%	81.7%	82.8%	85.9%
Delft	61.9%	63.1%	78.6%	64.4%	65.3%	79.6%
Amstelveen	63.8%	66.0%	82.9%	65.1%	66.3%	81.8%
Amsterdam	45.6%	51.1%	88.0%	50.4%	53.3%	87.5%
Den Haag	60.3%	61.7%	91.1%	63.4%	64.1%	90.3%

Table 2.4: Results as a percentage of maximum costs.

Figure 2.9: Solution of Delft, the chosen cascades.

all cabinets are activated is presented, which is the starting solution of the heuristic. For the Delft case with $t = 4$, for example, the heuristic found a better solution than the starting solution resulting in a saving of 36.9%, resulting in a score of 63.1%. The optimal solution of AIMMS is 1.2% better, resulting in a score of 61.9%. The random insertion method scored only 79.6%.

2.2.2 Clustering

The *cabinet clustering problem* is defined as follows: how to divide the cabinets in a number of clusters, where all cabinets in one cluster are connected in one circuit? When the telecom operator wants to connect the cabinets in a ring structure, he has to decide how to divide cabinets over a number of rings, taking into account a maximum number of customers per ring. In this section a model is developed and formulated and the results

2.2 Sequential approach

of extensive testing is shown. The method seems to be accurate and fast. Finally, the method is demonstrated on a real life case.

2.2.2.1 Problem definition

The starting point here is a collection of cabinets that have to be activated. These cabinets should be partitioned in groups, where all cabinets in one group have to be connected in one fibre ring. Mathematically this can be formulated as follows. Given is a set $D = \{x^i\}_{i=1}^m$ of m points in R^2, the location of the cabinets. These cabinets in have to be partitioned into k ($k \leq m$) groups each of which forms a ring. The cluster centres are centres of gravity, $C^1, C^2, \ldots, C^k$. We define for $i = 1, \ldots, m$ and $h = 1, \ldots, k$:

$$T_{i,h} = \begin{cases} 1 & \text{if data point } x_i \text{ is assigned to centre } C_h, \\ 0 & \text{otherwise.} \end{cases}$$

The cluster centres are calculated as follows, for $h = 1, \ldots, k$:

$$C^h = \frac{\sum_{i=1}^m T_{i,h} x^i}{\sum_{i=1}^m T_{i,h}}. \tag{2.7}$$

The problem that has to be solved is:

$$\min \sum_{i=1}^m \sum_{h=1}^k T_{i,h}(\|x^i - C^h\|_2^2). \tag{2.8}$$

This means that we look for the T that minimizes the sum of the quadratic distances, expressed in the sum of the squared 2-norm distance:

$$\|x\|_2 := \left(x_1^2 + x_2^2 + \cdots + x_n^2\right)^{\frac{1}{2}}. \tag{2.9}$$

In practice in this problem each ring has a maximum number of households that can be connected. This is modeled by assuming each cabinet has a weight (a natural number) u_i and there is a limit τ to the total weight on one ring. This gives the constraint:

$$\sum_{i=1}^m T_{i,h} u_i \leq \tau \quad (h = 1, \ldots, k). \tag{2.10}$$

Each point is assigned to only one ring:

$$\sum_{h=1}^k T_{i,h} = 1 \quad (i = 1, \ldots, m), \tag{2.11}$$

$$T_{i,h} \in \{0, 1\} \quad (i = 1, \ldots, m \quad h = 1, \ldots, k). \tag{2.12}$$

This problem is a *clustering problem*, more specifically 'Centroid-based' clustering. In centroid-based clustering, a centre point for every cluster is created, and points (cabinets in this case) are appointed to the closest centre point, in which the quadratic distance is minimized. If a number of k clusters is searched for, the method is called 'K-means' clustering. In the problem described above there is a maximum number of cabinets that can be placed in one circuit. This gives a constrained K-means clustering problem. In the literature much has been published about the unconstrained version of the problem, 'Centroid-based' or 'K-means' clustering, mostly in the applications of data-clustering. The name goes back to 1967, where MacQueen presents the problem in his article [101]. An nice overview of 50 years of 'K-means' clustering was presented by Jain [70]. That this optimization problem is NP-hard was proven by Aloise et al. [5]. Usually these problems are therefore solved via an approximation algorithm. In special cases the problem can be solved in polynomial time, see for example Inaba et al. [67]. For other cases a known method is Lloyd's algorithm [97]. However, this method is not guaranteed to find a global optimum, but usually finds a local optimum. Bradley et al. [16] create an extension to this algorithm to handle constrained K-Means Clustering. The constraint here is a minimum number of items in a cluster. In the cabinet clustering problem there is a maximum number of items in a cluster. We call this the max-constrained K-Means Clustering Problem.

Theorem 2.2.2. *The max-constrained K-Means Clustering problem is NP-hard.*

Proof. Setting $\tau = m$, the total number of points to be clustered, results in the standard K-Means Clustering Problem as defined and proven NP-hard in Aloise et al. [5]. □

2.2.2.2 Solution

In this section first the approaches by Lloyd [97] and Bradley et al. [16] are presented for later reference.

Algorithm 2 Lloyd's based on [97].

Require: A random distribution of the cabinets T^0. This results in an initial value of C^0.
Ensure: A set cluster points $C^{t,1}, \ldots, C^{t,k}$ that minimizes the distance of squared distances.
1: **for** $i = 1, \ldots, m$ **do**
2:    assign x_i to k such that centre $C^{k,t}$ is nearest to x_i;
3: **end for**
4: **for** $h = 1 \ldots, k$ **do**
5:    **if** $\sum_{i=1}^{m} T_{i,h}^t > 0$ **then**
6:       $C^{h,t+1} = \frac{\sum_{i=1}^{m} T_{i,h}^t x^i}{\sum_{i=1}^{m} T_{i,h}^t}$;
7:    **else**
8:       $C^{h,t+1} = C^{h,t}$;
9:    **end if**
10: **end for**
11: **if** $C^{h,t+1} = C^{h,t}, h = 1, \ldots, k$ **then**
12:    end algorithm;
13: **else**
14:    go to step 1;
15: **end if**

The algorithm of Lloyd is an iterative algorithm where T^t represents the assignment of the cabinets to the clusters in iteration $t \in \mathbb{N}$, resulting in cluster points $C^{t,1}, \ldots, C^{t,k}$

2.2 Sequential approach

in iteration t. The steps in the algorithm are presented in Algorithm 2. An often named disadvantage of this algorithm is that k needs to specified in advance. However, this is not a disadvantage in our clustering problem, here k is known beforehand. Also, it has been shown that the worst case running time of the algorithm is super-polynomial in the input size and the approximation found can be arbitrarily bad with respect to the objective function compared to the optimal clustering. Finding a good starting solution can be helpful here, as shown in [8]. This will be the case mainly in big data clustering problems, not in the limited problems that is discussed here. Another often mentioned disadvantage is that the algorithm tries to create clusters of approximately similar values. Here, this is not a problem either; on the contrary, it is a requirement. The cabinet cluster problem has an upper limit to the number of cabinets, or total weight of these cabinets, in a ring, as described in the constraint in Equation (2.10).

Bradley gives an extension to this algorithm to handle constrained K-Means Clustering. The constraint is a minimum number of items in a cluster, instead of a maximum as in the problem in this thesis, due to Equation (2.10). The algorithm of Bradley is presented in Algorithm 3.

Algorithm 3 Bradley's based on [16].

Require: A random distribution of the cabinets T^0. This results in an initial value of C^0.
Ensure: A set cluster points $C^{t,1}, \ldots, C^{t,k}$ that minimizes the distance of squared distances.
1: Let $T_{i,h}^t$ be a solution to the following linear program with $C^{h,t}$ fixed:

$$\min \sum_{i=1}^{m} \sum_{h=1}^{k} T_{i,h}(\|x^i - C^{h,t}\|_2^2),$$

subject to

$$\sum_{i=1}^{m} T_{i,h} \geq \tau_h \quad (h = 1, \ldots, k); \quad (*)$$

$$\sum_{h=1}^{k} T_{i,h} = 1 \quad (i = 1, \ldots, n);$$

2: **if** $\sum_{i=1}^{m} T_{i,h}^t > 0$ **then**
3: $\quad C^{h,t+1} = \frac{\sum_{i=1}^{m} T_{i,h}^t x^i}{\sum_{i=1}^{m} T_{i,h}^t}$;
4: **else**
5: $\quad C^{h,t+1} = C^{h,t} \quad h = 1, \ldots, k$;
6: **end if**
7: **if** $C^{h,t+1} = C^{h,t}, h = 1, \ldots, k$ **then**
8: $\quad$ end algorithm;
9: **else**
10: $\quad$ go to step 1;
11: **end if**

Replacing equation (*) in line 1 of the Algorithm by Equation (2.10) gives a solution method for our problem. However, as is shown later, the proposed heuristic performs better.

The algorithm proposed starts with Lloyd's algorithm. The found solution is adjusted such that no more than weight τ is placed in one ring, due to Equation (2.10), while Lloyd's algorithm does not have this check or constraint. The final step is trying to improve the solution using a 2-opt improvement, like originally presented in solving the Travelling Salesman Problem (TSP), see [32]. This is depicted in Figure 2.10 and in more detail in Algorithm 4. Some explanation, referring to the lines in the of the Algorithm, follows:

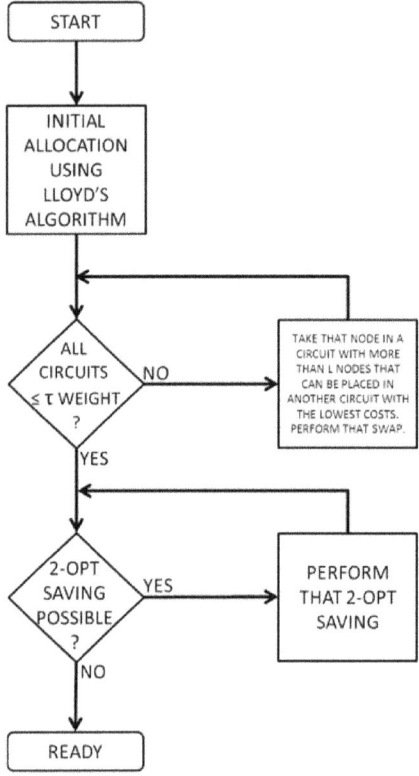

Figure 2.10: Flow diagram of the proposed heuristic.

1. Initial allocation according to Lloyd's algorithm, with $k = \lceil \sum_{i=1}^{n} u_i/\tau \rceil$. [5]

2. (Line 1-11) If a ring h is present with higher allocated weight than allowed, then choose the cabinet which can be moved to another with the least expenses. If no

---

[5]Note that the ratio between the values of u_i and k can lead to insolvable problems, as a simple example shows. Let us say that we have 5 cabinets with weight 4. We want to create rings with maximum weight 10. The definition of k says that 2 rings are needed ($\frac{4 \times 5}{10}$). Only we cannot divide the 5 cabinets over 2 rings, not violating the constraint. However, if the maximum value of u is small related to k and there is a big variation in number and place of the size of the cabinets (as in our practice) this problem never occurs.

2.2 Sequential approach

Algorithm 4 Clustering

Require: Initial allocation according to Lloyd's algorithm, with $k = \lceil \sum_{i=1}^{n} u_i / \tau \rceil$.
Ensure: A set cluster points $C^{t,1}, \ldots, C^{t,k}$ that minimizes the distance of squared distances.

1: **for** $h = 1, \ldots, k$ **do**
2:    **if** $\sum_{i=1}^{n} T_{i,h} u_i > \tau$ **then**
3:       $(i, l) = \arg\min_{\{(i|T_{i,h}=1), l\}} \|x^i - C^l\|^2$ under the constraint $\sum_{j=1}^{m} T_{j,l} u_j \leq \tau - u_i$;
4:       **if** (i,l) = empty **then**
5:          $((i,h), (i', h')) = \arg\min_{\{(i|T_{i,h}=1), (i'|T_{i',h'}=1)\}} \|x^i - C^{h'}\|_2^2 + \|x^{i'} - C^h\|_2^2$ under the constraints: $\sum_j (T_{j,h'} u_j) - u_{i'} + u_i \leq \tau$ and $\sum_j (T_{j,h} u_j) - u_i + u_{i'} \leq \tau$;
6:          $T_{i,h} = 0$; $T_{i,h'} = 1$; $T_{i',h'} = 0$; $T_{i',h} = 1$;
7:       **else**
8:          $T_{i,h} = 0$; $T_{i,l} = 1$;
9:       **end if**
10:    **end if**
11: **end for**
12: **for** every combination (i, i') where $T_{i,h} = 1$, $T_{i',h'} = 1$ and $h' \neq h$ **do**
13:    $T' = T$;
14:    $T'_{i,h'} = 1$, $T'_{i',h} = 1$, $T'_{i,h} = 0$ and $T'_{i',h'} = 0$.;
15:    calculate the centre points $C_T^1, \ldots, C_T^k$ based on solution T and $C_{T'}^1, \ldots, C_{T'}^k$ based on solution T';
16:    $S(T) = \sum_{j=1}^{m} \sum_{l=1}^{k} T_{j,l} (\|x^j - C^l\|_2^2)$;
17:    $S(T') = \sum_{j=1}^{m} \sum_{l=1}^{k} T'_{j,l} (\|x^j - C^l\|_2^2)$;
18:    **if** $\sum_{j=1}^{m} (T'_{j,h} u_j) \leq \tau$ and $\sum_{j=1}^{m} (T'_{j,h'} u_j) \leq \tau$ and $S(T') < S(T)$ **then**
19:       $T := T'$;
20:    **end if**
21: **end for**
22: If no swap occurred stop, else go to step 11;

solution can be found to remove the constraint using a single swap, a double swap must be found.

3. (Line 12-21) Try to improve the solution by pair swapping until no improvement can be found any more. Determine for every combination of cabinets within two separate rings the current score (total sum of quadratic distances) and the score after exchanging these two combinations. If the exchange results in an improvement, apply this exchange to the solution.

4. (Line 22) If no swap occurred stop, else go to step 3a.

2.2.2.3 Evaluation

In this section the performance of the proposed heuristic is discussed. First the effect of the swapping operation (step 3 in the heuristic) is shown. Next the performance against the method of Bradley is presented and finally the application of the method to the case Amstelveen is presented.

2.2.2.3.1 Swapping operation

The swapping operation, step 3 in the proposed heuristic, results in a clear improvement of the solution and a quicker convergence to the best solution. To illustrate this, 1000

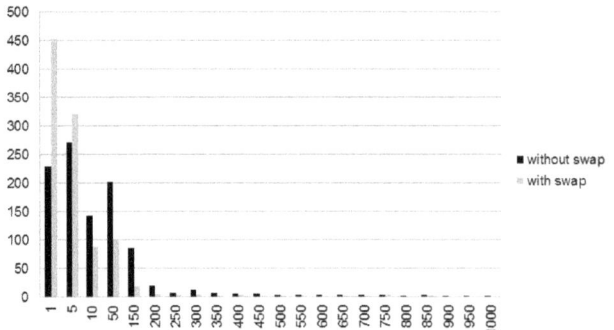

Figure 2.11: Iterations until best solution.

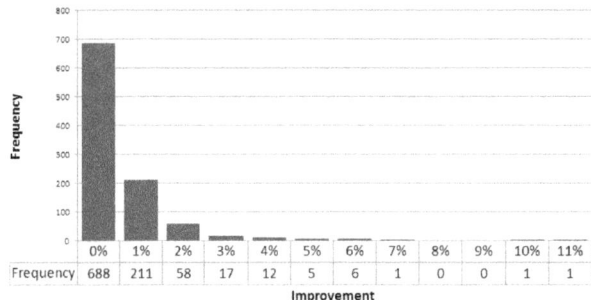

Figure 2.12: Performance improvement of swapping.

cases were generated each having 40 cabinets which need to be allocated to four circuits. The maximum number of cabinets per circuit is 10. The (x,y)-coordinates are arbitrarily drawn from the range (0,100) per cabinet. These cases were solved 1000 times, each time with an arbitrarily generated starting solution (like Lloyd's algorithm prescribes). For each case, the iteration for which the best solution is found and solution were saved. The algorithm was executed twice, with and without step 3 (the swap).

Figure 2.11 shows the number of iteration at which the best solution was found. The heuristic with swap finds the solution far quicker; in 452 of the 1000 cases the best solution is already found within the first iteration. The remaining 999 iterations offer no improvement. Without swap this is the case only 228 times.

The final solution found with swap is better than the heuristic without swap in 312 cases. In about 200 cases this improvement is 1%, in 60 cases 2% and in two cases even 10% and 11%, see Figure 2.12.

The question arises whether the swap method also results in an improvement of the classic clustering problem without the constraint. In the tested 1000 cases this shows to be only a minor improvement. In only a few cases the number of iterations is smaller and only in 38 cases an improvement of 1-2% is realized.

2.2 Sequential approach

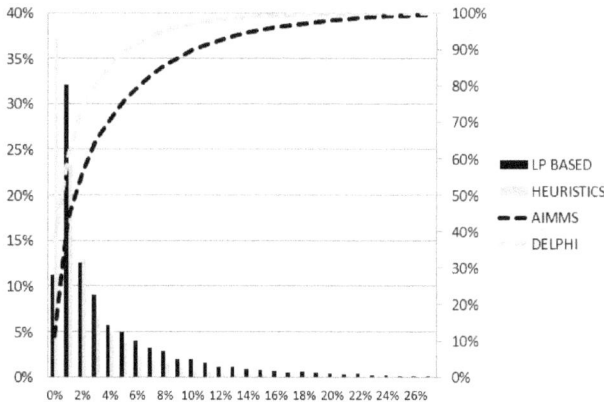

Figure 2.13: Deviation from optimal solution.

2.2.2.3.2 Numerical results

To evaluate the performance of the heuristic it was tested against the LP-based method proposed by Bradley [16]. To illustrate this, again 1000 cases were generated, in which there are 40 cabinets which need to be allocated to 4 circuits. The maximum number of cabinets per circuit is 10. The (x, y)-coordinates are arbitrarily drawn from the range (0,100) per cabinet. Per case we generated 20 starting solutions at random and looked at the solution found by each method for this starting solution. So we got 20,000 results per method. Per case the best solution found by both methods[6] and the relative deviation of the other solutions were saved. These results are depicted in Figure 2.13. We see that in 37% of all problems the heuristic found, starting with a arbitrary solution, the best solution against 11% in case of the LP-based method. The method of Bradley was implemented using AIMMS that solved the linear program using the solver CPLEX 12.4. The test version of the heuristic was implemented in Delphi.

Per case, having 20 starting solutions, the heuristic found in 81% the optimal solution against 33% in case of the LP-based method, as shown in Figure 2.14.

2.2.2.3.3 Real life case

Recall that that the calculation time is important in the real life use of the method presented. The implementation that was made in Matlab uses relatively much time for updating the cluster centres, especially due to the third step of the heuristic. Nevertheless, the calculation times are low. Take for example Amsterdam, where 235 activated cabinets of a total of 528 cabinets have to be clustered, which is one of the biggest central office of the Netherlands, it still can be calculated within 12 seconds. The results of all the cases are presented in Table 2.5.

Here the results of the Amstelveen case are presented. In Amstelveen there are 180 cabinets, belonging to 1 CO, serving 38,868 subscribers. The first step, activation, leaves 87 activated cabinets that have to be distributed over, in this case, 11 rings. A result of

---

[6]Note that this is not necessarily the global optimum, but only the best solution found by both methods.

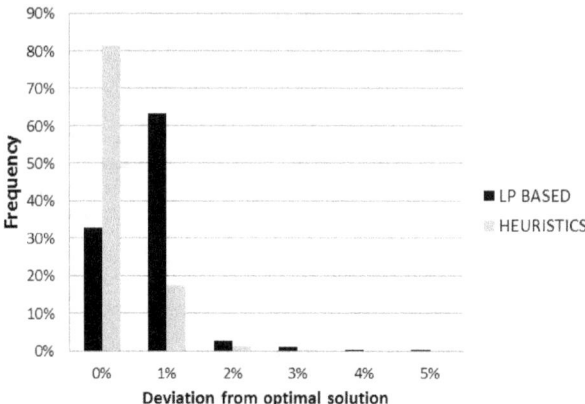

Figure 2.14: Performance of the heuristic.

City	Cabinets	Activated	Time
Venray	54	30	0.1s
Delft	130	60	0.8s
Amstelveen	180	87	0.9s
Amsterdam	528	212	11.4s
Den Haag	866	421	24.3s

Table 2.5: The results of the clustering for the cases.

the clustering of the Amstelveen case is shown in Figure 2.15, also showing the cascades resulting from solving the activation problem.

2.2.3 Routing

In this section the *routing problem* is formulated. This step also can be seen as the second, independent, step of the Cluster First - Route Second approach in a VRP. In the first step (the previous section) clusters of cabinets were created. Now all cabinets that belong to the same cluster have to be connected by edge disjoint paths. This problem is formulated and a solution method is proposed, based on solving a regular TSP and subsequently solving the conflicts that arise. The performance is evaluated again on the real life cases.

2.2.3.1 Problem definition

Let us assume that we have n cabinets that have to be connected. Underlying is a street or duct pattern to be used. This pattern can be modelled by points, of which some are connected by a street or duct. As those points do not have to be used in the solution, they can be seen as Steiner points. Assume that we have m of these Steiner points. Now a tour has to be found between all cabinets, using the Steiner points, such that the tour is as cheap as possible and no edge is used more than once. Both the cabinets as the

2.2 Sequential approach

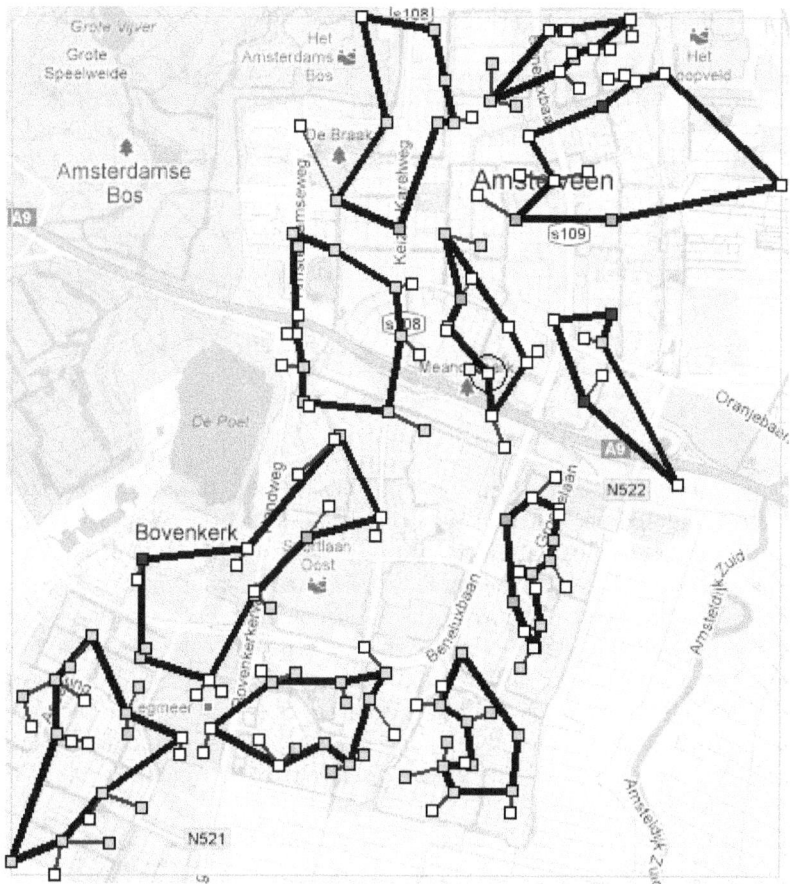

Figure 2.15: Solution of the Amstelveen case: assigned clusters.

Steiner points are called nodes in set N. The problem can be formulated as a IPP as follows. The decision variables for this problem are, for $i,j = 1, \ldots, n+m$;

$$x_{ij} = \begin{cases} 1 & \text{if node } i \text{ is linked to node } j, \\ 0 & \text{otherwise.} \end{cases}$$

$$z_j = \begin{cases} 1 & \text{if node } j \text{ is used,} \\ 0 & \text{otherwise.} \end{cases}$$

Now the objective is:

$$\min \sum_{i=1}^{n+m} \sum_{j=1}^{n+m} c_{ij} x_{ij},$$

where the constraints of this problem are:

$$z_i = 1 \quad (i = 1, \ldots, n), \tag{2.13}$$

$$x_{ij} \leq (z_i + z_j)/2 \quad (i, j = 1, \ldots, n + m) \quad i < j, \tag{2.14}$$

$$\sum_{j=1}^{n+m} x_{ij} + \sum_{j=1}^{n+m} x_{ji} = 2z_i \quad (i = 1, \ldots, n + m), \tag{2.15}$$

$$\sum_{i \in H} \sum_{j \in H} x_{ij} \leq \sum_{i \in H-l} z_i + 1 - z_t, \tag{2.16}$$

where

$$l \in H, \quad H \subset N, \quad |H| \geq 3, \quad t \in N - H,$$

$$x_{ij} \in \{0, 1\} \quad (i, j = 1, \ldots, n + m),$$

$$z_j \in \{0, 1\} \quad (j \in N).$$

These constraints make sure that all the cabinets are in the ring (2.13), no edges are used twice (2.15), only edges can be used to used nodes (2.14) and no sub-tours can occur (2.16). The sub-tour elimination constraint is based on Xu et al. [153], who also show that this problem is NP-hard. Xu et al. solve a similar problem with Tabu-search having, however, the ring structure only in the Steiner points. The target nodes (in our case the cabinets) are all connected to one of the Steiner points. Next to this, however the solution is presented as being fast, it is not fast enough for our goal. The problem solved in [153] has 300 target nodes and 300 Steiner nodes and is solved in seconds with a simple method to more than hundred seconds with a more complicated method. In our case the number of Steiner points will typically be around 300 times 300, much bigger than their problem.

Algorithm 5 Dijkstra(G,w,s) based on [30].

Require: A weighted directed graph $G = (V, A)$ with weights $w_{uv} \geq 0$ for each arc $(u, v) \in A$, a source vertex s.
Ensure: the cost $\delta(s, v)$ of the shortest path from source s to each $v \in V$ and a predecessor $\pi[v]$ in the shortest path for each $v \in V$.
1: **for** each vertex $v \in V$ **do**
2:    $d[v] := \infty$; {initialize shortest path cost for vertex v}
3:    $\pi[v] := NIL$; {initialize predecessor for vertex v}
4: **end for**
5: $d[s] := 0$; {shortest path from s to itself is 0}
6: $S := \emptyset$; {Initialize the set of processed vertices}
7: $Q := V$; {Initialize the set of vertices still to be processed}
8: **while** $Q \neq \emptyset$ **do**
9:    $u := EXTRACTMIN(Q)$; {Start processing the vertex u in Q with lowest $d[u]$}
10:    $S := S \cup \{u\}$;
11:    **for** each vertex $v \in Adj[u]$ **do**
12:      **if** $d[v] > d[u] + w(u, v)$ **then**
13:        $d[v] := d[u] + w[u, v]$; {relaxation operation}
14:        $\pi[v] := u$;
15:      **end if**
16:    **end for**
17: **end while**

2.2.3.2 Solution

To solve this problem the following heuristic is presented. We first create an initial solution by solving the underlying Travelling Salesman Problem. Here the requirement that no edges should be used more than once is not taken into account. With Dijkstra's algorithm [40, 30] all shortest paths between the cabinets that have to be connected are calculated and within that graph a minimal Hamilton circuit is created. This is calculated over all Steiner points. The used implementation of Dijkstra's algorithm is shown in Algorithm 5. For the Hamilton circuit the implemented method is: Brute force enumeration when number of cabinets smaller than ten, a combination of an *Arbitrary Insertion* method and 2-opt, as described by [65], otherwise. This gives an initial solution that is not necessarily feasible: multiple paths in the ring might use the same edges. To get a feasible solution we try to find two paths to substitute the paths with the double used edge: first find the shortest path from source s_1 to target t_1. Next, delete the edges used in this path from the graph and find the shortest path from source s_2 to target t_2 in this subgraph. Then reverse the procedure: find the shortest path from s_2 to t_2 (in the original graph), delete it from the graph and search for the shortest path from s_1 to t_1 in this subgraph. Then choose the cheapest option from the two. The whole procedure is given in more detail in Algorithm 6. Note that the while-loop (Lines 5-13) finishes in a finite number of steps. In fact, since each time the rest of the ring is fixed, it will finish in at most $R_r^C - 1$ steps. When this method, which is called the deletion method, fails to find a feasible solution, the whole algorithm finishes without finding a solution (Line 10). Assuming that a solution does exist, a remedy would be to take into account more than two conflicting paths at the same time, but no methods are known for this more general case.

Algorithm 6 Routing

Require: k clusters with cabinets: $C_1, ..., C_k$
Ensure: k rings $R_1^P, ..., R_k^P$ (and the $R_1^C, ..., R_k^C$), that are edge disjoint (not mutually) and their corresponding costs $cr_1, ..., cr_k$.
1: **for** $r = 1, ..., k$ **do**
2:    Compute the shortest path between all $s \in C_r$ using Dijkstra(G,w,s) for all $s \in C_r$;
3:    Solve a TSP for C_r to obtain R_r^C, the order of the cabinets in ring r, the paths of the ring R_r^P, and the cost cr_r of the ring;
4:    Create a list L which contains paths that use one or more of the same edges;
5:    **while** $L \neq \emptyset$ **do**
6:       Take the first conflict from the list and use the deletion method to fix the conflict. To avoid new conflicts, delete the rest of the ring from the graph;
7:       **if** the deletion method gives a feasible solution **then**
8:          Update R_r^P and cr_r;
9:       **else**
10:         STOP;
11:       **end if**
12:       Update the list L of conflicting paths in R_r^P;
13:    **end while**
14: **end for**

City	Cabinets	Activated	Calculation Time
Venray	54	30	2.4s
Delft	130	60	4.2s
Amstelveen	180	87	5.1s
Amsterdam	528	212	11.2s
Den Haag	866	421	21.0s

Table 2.6: The selected cases, calculation time for routing.

2.2.3.3 Evaluation

Solution was tested on the presented cases with an underlying Manhattan street pattern, not the real streets. The Steiner points are a grid of points with resolution of this pattern if 50 meter. The results are shown in Table 2.6. We see that even for the biggest CO area the 27 rings can be calculated within 12 seconds. The solution for Venray is depicted in Figure 2.16

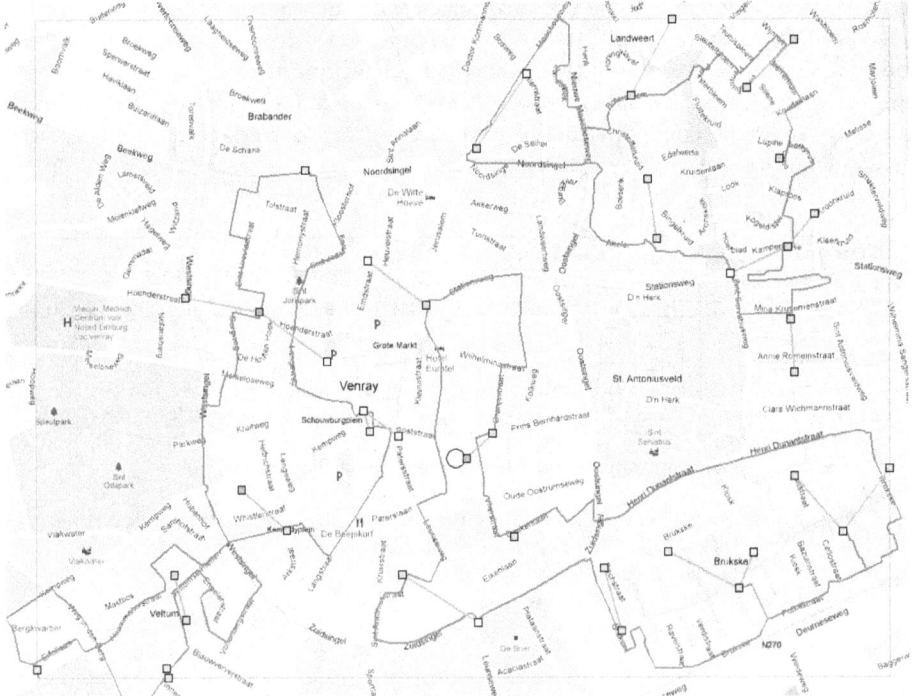

Figure 2.16: Solution of Venray: detailed routing.

2.3 Integrating clustering and routing

In this section the Edge Disjoint Circuits Problem (EDCP) that solves the clustering and routing problem together is presented. In the three step approach of the previous section those problems were solved sequentially, also known as the Cluster First, Route Second (CFRS) approach. The idea of the integral solution is that the solution will be better, while you use information of the routing in the clustering phase. For this approach a greedy insertion heuristic is proposed. After the construction of an initial solution, local search is used to improve the solution. Finally, the performance of the insertion heuristic is compared to the CFRS heuristic.

2.3.1 Problem definition

The EDCP can formally be described in the following way. Given is an undirected weighted graph $G = (V, E)$, where V denotes a set of vertices, representing the CO and the cabinets and E denotes a set of edges that connect the vertices. The set of cabinets is denoted by $SC \subset V$.

The input of the model is described by the following parameters:

q_i = number of clients served by cabinet i.
Cap = maximum number of customers that can be included in a ring.
a_{ij} = length of the edge from cabinet i to cabinet j; if there is no edge, $a_{ij} = 0$.
w_{ij} = the weight of the edge from cabinet i to cabinet j, which depends on the length of the edge and on whether the edge already has a duct available with free space.

The weight w_{ij} can be calculated given a subgraph $F = (V^F, E^F)$ of G. Contrary to G, F is unweighted and its edges are the ducts with available capacity. In other words, if $(i,j) \in E^F$ then this means that on this edge a duct with available capacity is present. Now introduce a binary parameter f_{ij} for which $f_{ij} = 1$ if $(i,j) \in E^F$, and $f_{ij} = 0$ otherwise. The weight w_{ij} of an edge $(i,j) \in E$ can then be calculated in the following way. If $f_{ij} = 0$, then $w_{ij} = a_{ij} \cdot \alpha$, where α is length unit cost of the fibre (including the digging, working, etc.). If $f_{ij} = 1$, then $w_{ij} = a_{ij} \cdot \alpha \cdot \beta$, where β is the cost factor reduction as a result of being able to use an existing pipe with available capacity.

The decision variable is denoted by R_r^C, a vector of ordered cabinets that are in ring r and by R_r^P the same ring in terms of edges and vertices of the original graph G. Then the problem can be formulated in the following way:

$$\min \sum_{r=1}^{k} \sum_{(i,j) \in R_r^P} w_{ij}, \qquad (2.17)$$

under the constraints:

$$CO \in R_r^C \quad (r = 1, \ldots, k), \qquad (2.18)$$

$$i \in \bigcup_{r=1}^{k} R_r^C \quad (i \in SC), \qquad (2.19)$$

$$\sum_{i \in SC \cap R_r^C} q_i \leq Cap \quad (r = 1, \ldots, k), \tag{2.20}$$

$$(i,j) \in E \text{ in } R_r^P \text{ at most once} \quad (r = 1, \ldots, k). \tag{2.21}$$

The constraints are explained by:

(2.18) The CO is included in each ring.
(2.19) Each cabinet or client is included in a ring.
(2.20) Maximum number of customers in ring not exceeded.
(2.21) Each ring is edge disjoint.

The problem studied in this section is a generalization of the Capacitated Vehicle Routing Problem (CVRP), which is shown to be NP-hard by Lenstra and Rinnooy Kan [94]. It is capacitated, since only a limited number of cabinets can be combined in a circuit. Additionally the underlying street pattern and the restriction that a ditch cannot be used twice has to be taken into account (restriction 2.21 above). This prevents applying the solution methods of the CVRP directly.

Theorem 2.3.1. *The EDCP is NP-hard.*

Proof. Setting $Cap = 1$ results in the standard Undirected Edge-Disjoint Paths Problem as defined and proven NP-hard in Andrews and Zhang [6]. □

2.3.2 Initial solution

In this section a method is outlined to find an initial solution for the EDCP of good quality where routing and clustering are solved simultaneously. It is important that the initial solution is of good quality, since, due to the edge disjointness restriction, local search techniques to improve the solution takes quite long for large instances.

One important restriction in the problem definition is the edge disjointness of a ring. A ring consists of shortest paths between cabinets that each consist of possibly several edges. Therefore, in this section relevant literature on the disjoint shortest paths literature is discussed.

Many papers have been published from the 1950's onwards that deal with all kinds of versions of disjoint path problems (see, e.g., the surveys in Korte et al. [85]). For some versions an algorithm is found that can solve the problem in polynomial time, whereas some other versions are proven to be NP-complete. The versions differ in which paths should be disjoint, in whether the number of disjoint paths is fixed, whether the paths need to be edge or node disjoint and the type of graph that is considered. Relatively few papers focus, however, on the optimization version of the problem (disjoint *shortest paths*) that is of interest for the research in this thesis. The literature on this disjoint *shortest* path problem is discussed next.

Early research efforts on the disjoint shortest paths problem include the work of Suurballe and Tarjan [140], who developed a polynomial algorithm for the problem to determine two disjoint shortest paths between a pair of vertices (see Figure 2.17). The idea behind the algorithm is very similar to Dijkstra's algorithm and it therefore computes two disjoint shortest paths from an origin vertex to all possible destination

2.3 Integrating clustering and routing

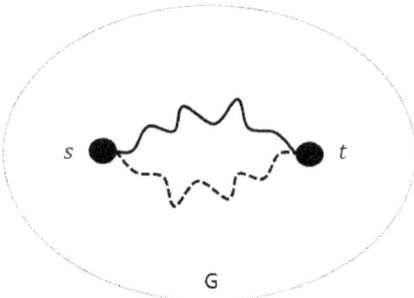

Figure 2.17: Edge disjoint shortest paths between a pair of vertices (s, t) as studied by [140].

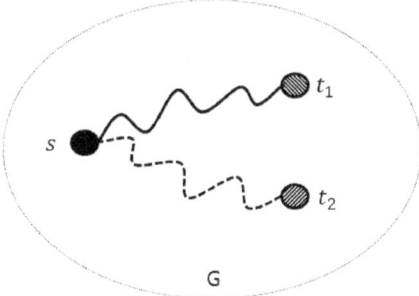

Figure 2.18: Edge disjoint shortest paths between a source s and two targets t_1 and t_2.

vertices (asymptotically) as fast as for only one destination vertex. Another problem is finding two disjoint shortest paths from a single source vertex to two destination vertices t_1 and t_2 (see Figure 2.18). This problem can be solved by introducing an artificial sink vertex t' that is connected to the two destination vertices t_1 and t_2. Then the problem is reduced to finding two disjoint paths from source s to target t'. This can be solved using the algorithm of Suurballe or by the perhaps more intuitive algorithm given by Bhandari [14]. Yang and Zheng [154] formulated two algorithms for this problem that have a slightly better asymptotic complexity. They are inspired by the work of [140].

Yet another, more general, problem is finding disjoint shortest paths between two pairs of vertices: one from source s_1 to target t_1 and one from source s_2 to target t_2 (see Figure 2.19). The objective that is of interest for the research here is the minimal sum of the path costs. Kobayashi and Sommer [79] look at two targets: minimum sum or minimum maximal length. For the general case of an undirected graph the state that no polynomial algorithm has been found nor has it been shown that the problem is NP-complete, also not for undirected planar graphs. Only for the special cases that the graph is undirected, planar and with sources and sinks incident to at most two faces of the graph, a polynomial algorithm is known. In most of the instances considered in this paper, however, this assumption of sources and sinks incident to at most two faces, does not hold. Note that the general directed version of this problem was shown to be

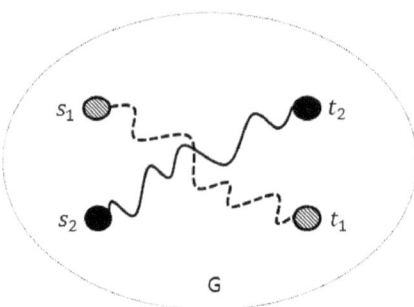

Figure 2.19: Edge disjoint shortest path pairs (s_1, t_1) and (s_2, t_2).

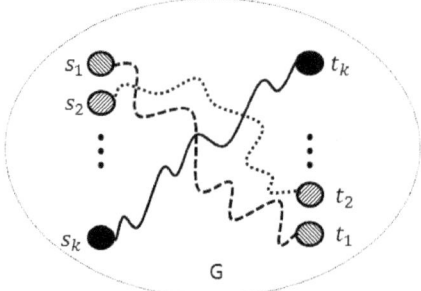

Figure 2.20: Edge disjoint shortest path pairs (s_1, t_1), (s_2, t_2) up to (s_k, t_k).

NP-hard by Fortune et al. [48].

Even more general is the problem to find k disjoint paths for the pairs $(s_1, t_1), (s_2, t_2)$ to (s_k, t_k), for which the sum of the path costs is minimal (see Figure 2.20). For $k = 2$ some special cases are known to be solvable in polynomial time. For $k = 3$ this is only the case, when all sources and sinks are incident to one face. All other special cases and $k > 3$ are still open research areas. Note that for k not fixed (so where it is still to determine how many of the pairs can be connected via a disjoint path) the problem is NP hard as shown by Karp [76].

The pseudo-code of our proposed heuristic is given in Algorithm 7. All the references to lines in the section below are to this algorithm. Some technical details are not included in this pseudo-code to improve the readability. In some steps more possible approaches are presented. Later in this section the different approaches are compared in performance. The output of the heuristic are k rings that are each edge disjoint. The objective is to minimize the sum of the ring costs. The algorithm starts with initializing the k rings by inserting one cabinet and the CO in each of them (k seeds need to be selected). Here k is a constant such that all cabinets can in fact be included in a ring, while the restriction of the maximum capacity of a ring is obeyed (see Line 2). In a later section the choice of k will be discussed further. To be able to start with a good initialization, first all the shortest paths between the cabinets are calculated using Dijkstra on the weight matrix W (see Line 1). Then the initialization can start (Lines

4-10) for which the following seed selection methods were studied:

1. *Random*: assign a cabinet randomly to each ring. This cabinet is then connected to the CO by two disjoint paths that are calculated by Suurballe's algorithm.

2. *Centroid with or without CO*: to obtain seeds to initialize the rings, use a clustering algorithm first and pick a seed from each cluster. The method of Section 2.2.2 can be used to group the cabinets in such a way that the sum of the Euclidean distances of each cabinet to the centroid of the cluster it is assigned to, is minimal. The CO can be forced to be present in each cluster or can be left out of the analysis. After clusters have been formed, the cabinet closest to the centroid of each cluster is chosen as a seed. In this way, both the capacities of the rings as well as the geographical spread is taken into account.

3. *Most spread*: this assigns seeds to the rings in a iterative procedure. For ring R_1^C add the CO and the cabinet furthest away (in terms of the shortest paths just computed). For ring R_2^C add the CO and the cabinet furthest away from the cabinet in ring R_1^C and the CO. For ring R_3^C add the CO and the cabinet furthest from R_1^C and R_2^C, etc. This is continued until all k rings contain the CO and one cabinet. In formula this is given by the expression in Line 5 of Algorithm 7. This initialization (that is also used in Naji-Azimi et al. [111]) ensures that the seeds are somewhat spread over the region with respect to the real costs in the graph.

After the seed cabinets are chosen, the actual rings can be built. In other words, for each of the rings $R_1^P, ..., R_k^P$, the actual disjoint paths (CO to cabinet and then back to CO) are calculated using Suurballe's algorithm (see Lines 4-10).

Now each ring has been initialized and the remaining cabinets (see Lines 11 and 12) should be included in the different rings in such a way that the total cost of the rings is as low as possible. This is done in the following way: insert the remaining cabinets one by one; each time looking at the cheapest ring and position to insert it. The order in which the cabinets need to be inserted can be determined in several ways. There are three possibilities:

1. *Random*: the cabinets are ordered in a random way and are inserted in this order.

2. *Disjoint insertion cost*: each time when deciding which cabinet to insert next, the costs is calculated to insert each unconnected cabinet in each ring in such a way that the edge disjointness constraint is satisfied. The simplest way to choose which cabinet to insert is to pick the one which can be inserted cheapest in a ring.

3. *Non-disjoint insertion cost*: each time the unconnected cabinet with the lowest non-disjoint insertion cost is inserted. These non-disjoint insertion costs are directly calculated from the shortest path costs that are calculated in Line 1 of Algorithm 7. The edge disjointness restriction is not necessarily satisfied and this option is therefore a compromise with regard to the previous option. It is less reliable, since non-disjoint cost can be quite deceiving, because of the inexpensive paths in the graphs that are studied. It gives, however, some indication and is therefore likely to perform better than a random order.

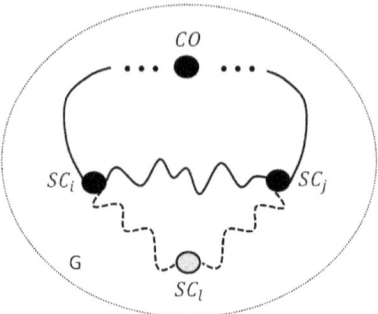

Figure 2.21: Illustration of the insertion of cabinet l in between cabinet i and cabinet j.

To decide which ring and position is actually cheapest for the insertion of a certain cabinet, for each ring and position the (disjoint) cost is computed (see Lines 13-21). An insertion of cabinet l in a ring r between cabinet i and cabinet j is illustrated in Figure 2.21. To compute the cost of an insertion the following procedure is used: remove the edges of the path from cabinet i to j from the ring. Then create a subgraph where the edges of this (disrupted) ring are deleted from the original graph. In this subgraph two disjoint paths are sought (from cabinet l to cabinet i and from cabinet l to cabinet j) in such a way that the total sum of the path costs is minimal. This is done using Suurballe's algorithm (see Lines 16-18). Removing the current ring from the graph that is used to search the new paths to connect cabinet l with the ring, ensures that no edges are used twice in a ring and the ring remains edge disjoint. Note that the insertion using Suurballe's algorithm ensures an optimal insertion, but fixing the rest of the ring makes the ring in the end not necessarily optimally routed. This is, however, the best that can be achieved with currently known disjoint shortest path algorithms. Given that cabinet l needs to be inserted next, the following steps are executed. Firstly, for each ring where adding cabinet l does not lead to exceeding the capacity, the non-disjoint insertion cost is computed for each position. Then a list is composed of all the just computed costs from low to high. Each element in the list corresponds to a ring and a position in that ring. Starting from the top of the list, the disjoint insertion costs are being computed. If the considered insertion is cheaper than the one found until that moment, the cheapest possibility is updated. In this way, an attempt is made to quickly find good disjoint solutions. Each time before considering the next option in the list, the non-disjoint cost is checked against the lowest disjoint insertion cost known until that moment. When the former is higher, there is no need to compute the disjoint shortest path costs anymore. In this way, a lot of computation time is saved. Another computation time saving technique is that before actually using Suurballe's algorithm, first the shortest paths (calculated in Line 1) are constructed from the predecessor relations to see if they are already coincidentally disjoint. If so, there is no need to use Suurballe's algorithm and computation time is saved.

2.3 Integrating clustering and routing

Algorithm 7 Insertion Heuristic

Require: A weighted undirected graph $G = (V, E)$ with weights $w_{ij} > 0$ for each edge $(i,j) \in E$, a set of cabinets $SC \subset V$, and a central office $CO \in V \setminus SC$. A parameter q_i for all $i \in SC$ indicating the number of customers it serves. A parameter Cap indicating the maximum number of customers in a ring.

Ensure: k rings $R_1^P, ..., R_k^P$ (and the $R_1^C, ..., R_k^C$), that are edge disjoint (not mutually) and their corresponding costs $cr_1, ..., cr_k$.

1: Compute the shortest path between all cabinets using Dijkstra(G,w,s) for all $s \in SC$. Save the cost in a matrix C of size $|SC|$ by $|SC|$.
2: $k := \lceil \sum_{i \in SC} q_i / Cap \rceil$ {A lower bound on the # of needed rings}
3: $R_0^C := \{CO\}$
4: **for** $r = 1, ..., k$ **do**
5:     $i := \arg\max_{i \in SC} C(\bigcup_{t=0}^{r-1} R_t^C, i)$. {'Most spread' or use one of the alternatives (see page 39).}
6:     $R_r^C := \{CO, i\}$.
7:     Run Suurballe(G,w,CO,i) to get P_1 and P_2 and the corresponding cost cp_1 and cp_2.
8:     Initialize R_r^P by combining the paths P_1 and P_2.
9:     $cr_r := cp_1 + cp_2$.
10: **end for**
11: **while** $SC \setminus \{R_1^C \cup ... \cup R_k^C\} \neq \emptyset$ **do**
12:     Select an $l \in SC \setminus \{R_1^C \cup ... \cup R_k^C\}$ according to some priority rule.
13:     **for** $r = 1, ..., k$ **do**
14:       **if** $\sum_{i \in SC \cap R_r^C} q_i + q_l \leq Cap$ **then**
15:         **for** $i = 1, ..., |R_r^C|$ **do**
16:           Introduce a new vertex t' and define $G^* = (V^*, E^*)$ by $V^* = V \cup \{t'\}$ and $w(R_r^C(i), t') := w(R_r^C(i+1), t') := w(t', R_r^C(i)) := w(t', R_r^C(i+1)) := \epsilon > 0$.
17:           Run Suurballe($G^* \setminus \{R_r^P \setminus P(i,j)\},w,t',l$), where $P(i,j)$ denotes the current path between i and j in the ring R_r^P. If a solution exists, denote by P_1 and P_2 the two disjoint paths and by cp_1 and cp_2 the corresponding costs. If no solution exists set cp_1 and cp_2 to a high number M.
18:           $cost(r, i) := cp_1 + cp_2 - c(i, j)$.
19:         **end for**
20:        **end if**
21:     **end for**
22:     Use the combination of $r \in \{1, ..., k\}$ and $i \in \{1, ..., |R_r^C|\}$ s.t. $cost(r, i)$ is minimal to adapt the rings:
23:     $R_r^C := \{R_r^C(1), ..., R_r^C(i), l, R_r^C(i+1), ..., R_r^C(end)\}$
24:     Insert l in R_r^P in between SC $R_r^P(i)$ and SC $R_r^P(j) := R_r^P(i+1)$ by deleting the path $P(i,j)$ and using P_1 and P_2 corresponding to the pair (r, i) instead.
25:     $cr_r := cr_r - cp(i,j) + cp_1 + cp_2$, where $cp(i,j)$ is the cost of the current path from (i,j) in ring r, cp_1 the cost of path P_1 and cp_2 the cost of path P_2.
26: **end while**

2.3.3 Improvement

After an initial solution is formed with one of the heuristics discussed in the previous section, local search can try to improve the solution. Important to note is that in this phase of the algorithm only changes are made that do not affect the feasibility of the solution. There are several options: some acting on a single ring, others acting on multiple rings simultaneously. To decrease the cost of a ring one could try to do the following neighborhood operations within one ring (see e.g. Kindervater and Savelsbergh [77]):

1. k−Exchanges, i.e. replace a set of k edges by another set of k edges. According to [77] in practical applications only 2-exchanges and 3-exchanges are relevant, since otherwise the computation time becomes too large.

2. Relocation, i.e. relocating a cabinet to another position in the ring (or more generally relocating a set of l-consecutive nodes).

Also neighborhoods for multiple rings can be defined. In this way cabinets that were assigned to a certain ring in the initial solution can be transferred to other rings. Among the possibilities are the following:

1. Exchange a cabinet from one ring with a cabinet from another ring.

2. Relocation, i.e. relocate a cabinet to another ring. It then has to be decided in which ring and in which position the cabinet is inserted. A cabinet will only be relocated to another ring if that ring has enough capacity left.

These different neighborhoods could be used within a metaheuristic, such as tabu search or simulated annealing. These metaheuristics are, however, only useful if savings of a certain neighborhood operation can be calculated very quickly. Unfortunately, this is not the case for the problem studied in this section; already for medium sized instances the computation of edge disjoint paths becomes too time intensive to do them as many times as preferable in a metaheuristic. Instead, a hill climbing approach is chosen. In hill climbing an initial solution is improved by neighborhood moves until no further improvements can be made.

The neighborhood moves relocation and exchange of cabinets between rings, can be implemented in various ways. Two options are investigated:

1. A consecutive implementation of relocation and exchanges.

2. An implementation that integrates the relocation and exchange in one.

The first option is particularly interesting when the instances are larger and not much time is available for the local search. Neighborhood searches for relocation moves are namely much faster than those for exchange moves. The second approach is likely to be able to find solutions of better quality than the first approach, provided that the alternation between relocations and exchanges is done in a sound manner.

To decide which of the two types of moves is performed, the following logic is used: candidate cabinet l_1 is chosen and for this cabinet the cheapest ring and position in all other rings with available capacity is calculated (disjoint insertion cost). Also the cheapest exchange with another cabinet l_2 from another ring is calculated for the rings where it was not possible to relocate cabinet l_1 to because of capacity restrictions. The rationale behind this idea is that it saves a lot of computation time, while at the same time the rings that are not considered for exchanges are still covered by the relocation moves.

Similar to the Insertion Heuristic, the decision remains in which order to consider the cabinets. A random order can be chosen, but a more sensible option is ordering the cabinets according to the sum of the costs of the connections to their neighbors in the ring. The ordered list of cabinets is run through and as soon as for a cabinet either a cost decreasing relocation move or a cost decreasing exchange move is found, the move is executed. An alternative would be to search the entire neighborhood and select the move that reduces the cost the most. This may lead to a better solution or less neighborhood searches. Searching the entire neighborhood each time would, however, take too much computation time. It is important to improve the solution as quickly as possible, since the maximal time reserved for improvements is, especially for larger instances, quite limited and the local optimum will often not be reached before this time bound is exceeded.

Finally, note that the local search can stop as a result of two situations:

1. No improvements can be found anymore. In other words, a local optimum is reached.

2. The computation time reaches its maximum (e.g. 5 minutes).

The second situation will occur for the larger instances, whereas for the small instances often a local optimum is reached earlier.

2.3.4 Evaluation

In this section the results of the thorough testing are presented. Here the different options within the insertion method under various assumptions are compared and the insertion method is compared to the CFRS method of Sections 2.2.2 and 2.2.3. First the test instances used are presented.

2.3.4.1 Test instances

The focus in this section is on theoretical instances that are based on a rectangular grid. These grids are, however, made in such a way that the essential characteristics of real world instances regarding the cost patterns are present. The grid defines a graph as follows: let the nodes be the points on the grid where two lines cross and let the edges be the lines connecting each node to its neighbor. The cost of each edge is chosen randomly (using an uniform distribution within some specified bounds). Next, part of the edges of the graph are reduced in cost to represent the availability of ducts with available capacity. Different patterns of ducts with reduced cost are studied in this section. One of the duct patterns studied has ducts in paths through the graph. In real world instances, it is this type of duct patterns that will often be present. Ducts may namely have been placed for other networks, such as the core network of mobile communications, the core network of the broadband internet or fibre cables going to business areas, etc. The instance given in Figure 2.22 is an example of such an instance where ducts are available on paths in the graph. Another possible duct pattern is where ducts are available on a path from the CO to each cabinet. In other words, a tree network of ducts is available. A major part of this network can be used to construct rings, so it is relatively inexpensive to create rings to ensure a high reliability. This type of duct pattern is closely related to the one studied by Kalsch et al. [75]. Their methodology is specifically targeted at embedding ring structures in large fibre networks.

Next, part of the vertices of the graph are chosen randomly to have a special meaning. The most centrally located one in the grid will represent the CO, while the others will function as cabinets. For each cabinet the number of customers is chosen randomly according to some specified distribution. An example of a (small) grid instance is given in Figure 2.22. On the edges, the costs of the edge is indicated. Moreover, edges on which a duct is available with free capacity are drawn as solid lines, whereas when there is no duct available, but digging is allowed, a dotted line is used. The CO is marked by a circle and numbered one. The other cabinets are numbered from 2 to 15.

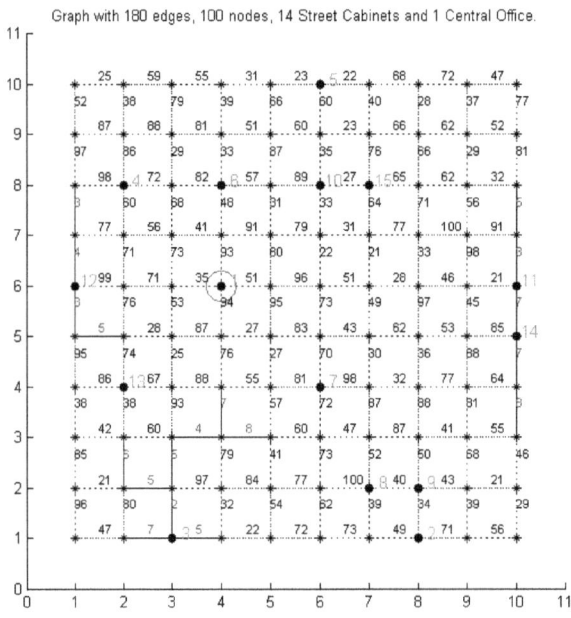

Figure 2.22: An example grid instance of size 10 by 10 with 14 cabinets and a CO. Ducts are available on random paths in the graph.

Three different computers were used in the calculation, for practical reasons. They varied a little in their speed and therefore in each table the used computer is mentioned. The specifications of the different computers are:

1. PC A: Intel Core i5 CPU M430 2.27Ghz; 6GB RAM; NVIDIA GeForce GT 320M

3.7GB.

2. PC B: Intel Core i3 CPU M330 2.13GHz; 4GB RAM; Intel HD Graphics 1.6GB.

3. PC C: Intel Core i5-2520M CPU 2.50GHz; 4GB RAM; Intel HD Graphics 1.6GB.

2.3.4.2 Choices within the insertion method

In Section 2.3.2 different options were described to select seeds to initialize the rings, the insertion order and the number of rings. We look here at the performance of the various choices.

2.3.4.2.1 Initialization of the rings in the insertion algorithm

The first cabinet in each ring can be chosen randomly, based on the clustering algorithm of Section 2.2.2 or by choosing cabinets in such a way that they are most spread over the grid (according to cost). The second option has two variants: including the CO in the clustering or leave it out. These four initialization types are tested on instances differing in size, duct patterns, customer distributions, etc. To be able to draw a reliable conclusion random instances are drawn many times for each instance specification. Note that only the comparison within a line of a table is relevant. From Tables 2.7 to 2.10 we learn that none of the four options outperforms the others for *all* random instances. For all instances (except one) of the type 'random inexpensive paths' the 'most spread' seed selection method clearly performs the best. Both the average cost and the proportion of all replications it finds the lowest cost consequently favor this type of initialization. An exception is the large grid (75 by 75) with no cheap ducts available at all. There 'centroid without CO', surprisingly, performs the best. In most instances from practice, cheap ducts are in fact available, but their amount differs per region. It is, however, unlikely that no ducts are available at all. Since the seed selection 'most spread' performs the best for all the from a practical perspective 'interesting' instance types, it is chosen henceforth as standard for instances of the type 'inexpensive random paths'.

For the instances where there is an inexpensive path from the CO to each of the cabinets, the seed selection by using the 'Centroid without CO' has the best performance with respect to the average cost and '% best'. Only for the smallest instances, 'most spread' performs better than using the clustering algorithm of Section 2.2.2 to get seeds for the rings. A little more computation time is needed to get the seeds, but this extra computation time is still reasonably small and it can be worth it to get a better solution quality. Therefore, for instances where all cabinets can be reached via an inexpensive path, seed selection is, henceforth, done using the 'centroid without CO' method.

However, given our expectations about the real life cases, the option 'most spread' was chosen as default in our testing.

2.3.4.2.2 Insertion order

Three different options were given concerning the insertion order of the cabinets: randomly, based on non-disjoint insertion cost and based on disjoint insertion cost. These are the three basic options, which are tested on their performance. In the Tables 2.11 to 2.13 the results are shown for various instance sizes and duct patterns. All times are in

Duct Pattern	Random			Centroid with CO			Centroid without CO			Most spread		
	best	cost	time	best	cost	time	best	cost	time	best	cost	time
10 cheap paths	24%	1,700	0.14	29%	1,686	0.15	9%	1,675	0.15	39%	1,669	0.13
5 cheap paths	19%	2,209	0.11	31%	2,167	0.13	9%	2,153	0.12	42%	2,132	0.10
0 cheap paths	20%	2,953	0.04	31%	2,894	0.06	10%	2,867	0.06	39%	2,844	0.04
CO to each Cabinet	22%	2,210	0.09	30%	2,167	0.11	10%	2,161	0.11	38%	2,148	0.09

Table 2.7: Test results for seed selection method for a grid of size 10 by 10 with 15 cabinets for various duct patterns. Other parameters: cap=2500, cust=DU(100,200,300,400,500), cost reduction pipes: 90%, lengthedge=U(20,100). Computer 1, 1000 replications.

Duct Pattern	Random			Centroid with CO			Centroid without CO			Most spread		
	best	cost	time	best	cost	time	best	cost	time	best	cost	time
30 cheap paths	14%	9,038	3.54	19%	8,902	3.73	30%	8,769	3.71	38%	8,745	3.32
15 cheap paths	12%	11,775	3.33	19%	11,505	3.45	30%	11,350	3.39	40%	11,305	3.07
0 cheap paths	7%	20,113	0.84	19%	19,411	1.15	24%	19,241	1.15	51%	18,755	0.84
CO to each cabinet	12%	7,615	6.32	29%	7,244	6.50	42%	7,136	6.58	18%	7,535	6.48

Table 2.8: Test results for seed selection method for a grid of size 30 by 30 with 50 cabinets for various duct patterns. Other parameters: cap=8, cust=1, cost reduction pipes: 90%, lengthedge=U(20,100). Computer 1, 1000 replications.

Duct Pattern	Random			Centroid with CO			Centroid without CO			Most spread		
	best	cost	time	best	cost	time	best	cost	time	best	cost	time
50 cheap paths	7%	21,785	14.16	17%	21,521	15.80	20%	21,408	15.47	57%	20,871	14.11
25 cheap paths	10%	28,284	15.20	18%	27,848	16.53	27%	27,715	15.65	46%	27,229	14.62
0 cheap paths	6%	56,314	4.07	28%	54,511	6.53	27%	54,486	6.48	40%	53,849	4.11
CO to each cabinet	11%	18,942	27.66	27%	18,282	27.80	33%	18,073	27.17	29%	18,383	28.94

Table 2.9: Test results seed for selection method for a grid of size 50 by 50 with 100 cabinets for various duct patterns. Other parameters: cap=2500, cust=DU(100,200,300,400,500), cost reduction pipes: 90%, lengthedge=U(20,100). Computer 1, 100 replications

Duct Pattern	Random			Centroid with CO			Centroid without CO			Most spread		
	best	cost	time	best	cost	time	best	cost	time	best	cost	time
75 cheap paths	7%	70,525	104.58	11%	69,792	148.86	32%	69,022	147.50	50%	68,347	113.75
40 cheap paths	8%	86,140	126.08	19%	85,279	169.50	30%	84,583	164.25	43%	84,471	132.70
0 cheap paths	6%	206,289	36.41	49%	199,757	85.04	28%	201,459	83.78	17%	203,108	36.65
CO to each cabinet	7%	55,012	194.16	19%	53,408	226.35	28%	53,102	228.39	46%	52,823	219.94

Table 2.10: Test results for seed selection method for a grid of size 75 by 75 with 300 cabinets for various duct patterns. Other parameters: cap=8, cust=1, cost reduction pipes: 90%, lengthedge=U(20,100). Computer 1, 100 replications.

seconds. These tables show that the larger the instance (both in terms of grid sizes as in the number of cabinets), the bigger the proportion of replications where the disjoint cost order leads to the lowest average cost ('% best'). For the instances of size 25 by 25 and 50 by 50, the percentage of times the disjoint cost order was the best over the different replications lies around 50% to 60%, whereas for the instances of size 75 by 75

2.3 Integrating clustering and routing

is it closer to 90%.

The random order has the highest average cost for all tested instances. It is better to insert the cabinet with the lowest non-disjoint insertion cost each time. This does take a little more computation time, but leads to a lower average cost. Finally, the disjoint cost order has the lowest average cost, but the computation time is considerably larger. For the larger instances of size 75 by 75 it can take up to an hour to come to an initial solution, while the non-disjoint cost order is only using 2 to 3 minutes. The disjoint cost order does not satisfy the requirements set in the problem definition. Given our expectations about the real life cases, the option 'non-disjoint insertion cost' was chosen as default in the testing.

Duct Pattern	Random order			Non-disjoint cost order			Disjoint cost order		
	best	cost	time	best	cost	time	best	cost	time
25 cheap paths	23%	9,954	1.78	26%	9,746	1.89	51%	9,576	34.00
13 cheap paths	9%	12,214	1.77	27%	11,927	1.63	64%	11,656	33.47
0 cheap paths	19%	17,420	0.86	46%	16,891	0.90	35%	16,830	22.28
CO to each cabinet	17%	8,487	3.26	23%	8,201	3.70	60%	7,919	50.41

Table 2.11: Test results for insertion order for a grid of size 25 by 25 with 60 cabinets for various duct patterns. Other parameters: cap=2000, cust=DU(100,200,300,400,500), cost reduction pipes: 80%, lengthedge=U(20,100). Computer 2, 100 replications

Duct Pattern	Random order			Non-disjoint cost order			Disjoint cost order		
	best	cost	time	best	cost	time	best	cost	time
50 cheap paths	19%	30,749	6.10	28%	30,253	6.72	53%	29,910	197.28
30 cheap paths	18%	34,622	6.85	26%	34,098	7.88	56%	33,748	219.02
13 cheap paths	21%	42,266	5.92	28%	41,578	6.78	51%	40,835	195.02
CO to each cabinet	14%	28,033	8.71	20%	27,972	10.13	66%	26,711	260.80

Table 2.12: Test results insertion order for a grid of size 50 by 50 with 100 cabinets for various duct patterns. Other parameters: cap=2500, cust=DU(100,200,300,400,500), cost reduction pipes: 70%, lengthedge=U(20,100). Computer 2, 100 replications

Duct Pattern	Random order			Non-disjoint cost order			Disjoint cost order		
	best	cost	time	best	cost	time	best	cost	time
75 cheap paths	0%	54,252	128.12	8%	53,111	153.55	92%	50,421	3492.98
40 cheap paths	4%	84,454	70.18	4%	83,761	88.52	92%	80,326	2602.27
0 cheap paths	0%	228,577	20.78	36%	203,737	33.96	64%	202,778	2063.91
CO to each cabinet	8%	54,002	134.93	3%	53,156	163.61	89%	50,492	3762.66

Table 2.13: Test results for insertion order for a grid of size 75 by 75 with 300 cabinets for various duct patterns. Other parameters: cap=8, cust=1, cost reduction pipes: 90%, lengthedge=U(20,100). Computer 3, 25 replications

2.3.4.2.3 Influence of the initial number of rings

An interesting question raised is: can the cost of the Insertion Heuristic be lowered (on average) if a certain percentage of extra rings is added from the start to the lower bound

$$k = \lceil \sum_{i \in SC} q_i / Cap \rceil?$$

We observed that the average costs of the different number of rings are pretty close to each other. Still, it is worth starting with a high number of initial rings as is visible from the tables. Instances with different characteristics are used to be able to get a good overview. It takes, however, a lot of time to compute initial solutions for so many different initial rings. This makes it hard to come to a conclusion that is very reliable. An attempt is made, but much more testing would have been preferable; both concerning instance types as the number of replications. The results give, however, some indication. When little time is available for the local search, the best results are obtained if a certain percentage of extra number of rings is chosen. Based on the testing in this section, the optimal percentage of extra rings is estimated to be around 14%. However, when much computation time is available, preliminary tests show that it might be wiser to choose no (or very little) extra rings at the start.

2.3.4.3 Comparison of the two methods

The most interesting part of the testing of the Insertion Heuristic is the comparison with the CFRS Heuristic based on Sections 2.2.2 and 2.2.3. These test results can be used to answer the question: does integrating the clustering and disjoint routing into one approach give better solutions than a separated approach? Before turning to the test results, note that it is almost impossible to draw a general conclusion on this question. Many options and tuning parameters are present in each of the approaches. What can be compared are the results of the CFRS Heuristic as implemented and the Insertion Heuristic with the chosen options.

2.3.4.3.1 Initial solution

First the initial solutions of both heuristics are compared. The analysis starts with the results of the instances where ducts are available as random paths through the graph. Afterwards, the test results for the instances where ducts are available from the CO to each of the cabinets, are treated.

The performance of the Insertion Heuristic is quite good compared to the CFRS Heuristic, however this heuristic takes less time, 40% less for bigger instances. When looking at the percentages in Tables 2.14 to 2.17, it can be concluded that the Insertion Heuristic always finds a pretty good solution compared to the CFRS Heuristic. Only for a few instances the performance of the CFRS Heuristic is better (in terms of average cost) than that of the Insertion Heuristic. The difference is, however, not that large: only 1% to 3% and they occur in medium-sized instances where no or a few cheap ducts are available.

The instances where an inexpensive path is available from the CO to each of the cabinets give very clear results: the Insertion Heuristic performs the best, both in terms of average cost as in the proportion of the replications it gives the best solution.

2.3 Integrating clustering and routing

Duct Pattern	Insertion Heuristic						CFRS Heuristic						no sol
	initial solution			after local search			initial solution			after local search			
	best	cost	time	best	cost	time	best	cost	time	best	cost	time	
20 cheap paths	58%	10,073	0.70	59%	9,217	80.76	42%	10,302	0.73	41%	9,288	101.30	2%
10 cheap paths	56%	11,062	0.66	56%	10,117	74.99	44%	11,164	0.73	44%	10,194	98.03	2%
0 cheap paths	58%	12,841	0.59	58%	11,773	61.67	42%	12,650	0.77	42%	11,873	72.84	3%
CO to each cabinet	62%	9,025	1.19	62%	8,082	83.09	38%	9,273	0.75	38%	8,210	129.11	2%

Table 2.14: Test results of the comparison between the Insertion Heuristic and the CFRS Heuristic for grid of size 20 by 20 with 50 cabinets for various duct patterns. Other parameters: cap=8, cust=1, cost reduction pipes: 50%, lengthedge=U(20,100). Computer 2, 100 replications. Max time local search: 5 min.

Duct Pattern	Insertion Heuristic						CFRS Heuristic						no sol
	initial solution			after local search			initial solution			after local search			
	best	cost	time	best	cost	time	best	cost	time	best	cost	time	
50 cheap paths	61%	34,370	3.74	63%	31,345	301.25	39%	34,827	2.61	37%	31,794	301.69	0%
25 cheap paths	64%	37,979	3.78	62%	34,706	301.12	36%	38,660	2.69	38%	35,084	300.79	0%
0 cheap paths	34%	53,688	2.61	45%	48,087	300.91	66%	52,312	2.44	55%	47,744	300.91	0%
CO to each cabinet	57%	32,458	4.99	72%	28,443	301.43	43%	32,694	2.78	28%	29,326	302.15	1%

Table 2.15: Test results of the comparison between the Insertion Heuristic and the CFRS Heuristic for grid of size 50 by 50 with 100 cabinets for various duct patterns. Other parameters: cap=2500, cust=DU(100,200,300,400,500), cost reduction pipes: 60%, lengthedge=U(20,100). Computer 3, 100 replications. Max time local search: 5 min.

Duct Pattern	Insertion Heuristic						CFRS Heuristic						no sol
	initial solution			after local search			initial solution			after local search			
	best	cost	time	best	cost	time	best	cost	time	best	cost	time	
60 cheap paths	100%	80,231	48.79	100%	77,829	309.82	0%	90,811	35.19	0%	86,764	305.17	1%
30 cheap paths	100%	93,156	53.96	100%	90,243	310.75	0%	107,376	34.68	0%	102,596	307.60	3%
10 cheap paths	100%	124,093	51.88	100%	120,282	308.32	0%	137,148	34.69	0%	132,483	304.08	1%
CO to each cabinet	100%	69,208	67.34	100%	65,972	311.72	0%	82,791	35.35	0%	77,721	306.09	1%

Table 2.16: Test results of the comparison between the Insertion Heuristic and the CFRS Heuristic for grid of size 70 by 70 with 250 cabinets for various duct patterns. Other parameters: cap=2500, cust=DU(100,200,300,400,500), cost reduction pipes: 75%, lengthedge=U(20,100). Computer 1, 100 replications. Max time local search: 5 min.

2.3.4.3.2 Solution after local search

After analysing the results of the initial solutions, now the effects of applying the local search are investigated. They are also depicted in Tables 2.14 to 2.17. Note that the computation time of local search was restricted to 5 minutes, since the practical considerations require the total computation time to be in the order of magnitude of minutes. Only for the smallest two instances the local search stopped in less than 5 minutes because a local minimum was reached. In all other instances, it was the maximal time which lead to the end of the local search.

Since the local search is often aborted before the local minimum is reached, the initial solution is very important. The results show that a lower cost of the initial solution leads

Duct Pattern	Insertion Heuristic						CFRS Heuristic						
	initial solution			after local search			initial solution			after local search			
	best	cost	time	best	cost	time	best	cost	time	best	cost	time	no sol
100 cheap paths	100%	148,182	657.84	100%	147,138	364.85	0%	187,148	399.81	0%	184,478	344.91	0%
50 cheap paths	100%	181,754	738.17	100%	180,383	374.91	0%	243,458	393.29	0%	239,713	349.01	0%
25 cheap paths	100%	225,423	727.83	100%	223,771	361.00	0%	309,579	395.22	0%	305,582	346.23	2%

Table 2.17: Test results of the comparison between the Insertion Heuristic and the CFRS Heuristic for grid of size 100 by 100 with 600 cabinets for various duct patterns. Other parameters: cap=2500, cust=DU(100,200,300,400,500), cost reduction pipes: 90%, lengthedge=U(20,100). Computer 2, 100 replications. Max time local search: 5 min.

to a lower cost of the local search. The local search gives an improvement of the initial solution of 10-15%.

2.4 Summary

In this chapter an approach was presented for the planning of FttCab. First a method based on three separate steps was proposed. The first problem was the cabinet activation problem: How can be decided which cabinets have to be activated to provide a certain level of clients within the VDSL range against minimal costs. The problem was defined, and a heuristic approach was developed and benchmarked to the optimal solution of the IPP, found by AIMMS/CPLEX on real cases in The Netherlands. The heuristic seemed to work good and very fast.

The second problem was the clustering problem: How can we divide access points, cabinets in our example, over a number of circuits, taking into account a maximum weight per circuit. To solve this, a heuristic was presented which uses Lloyd's algorithm, and added a re-clustering method to stay under the maximum weight and a 2-opt improvement method. Finally the results of our extensive testing on the effect of the swapping operation, step 3 in the heuristic on the total heuristic against the method of Bradley have been presented. We saw that the method is accurate and fast.

The third problem was the routing problem: How to connect the Clusters of the second problem with each other? An approach was presented in which was started with solving a general TSP and then repairing the paths in the ring that use the same edges. The problem was solved quickly by this approach.

Finally, the second and third problem together have been solved, introducing the EDCP. The questing here was: Does integrating the clustering and disjoint routing into one approach, named the CFRS Heuristic, give better solutions than a separated approach? In this respect a greedy insertion heuristic was developed to find an initial solution. Subsequently, this solution was improved using local search, until either the maximum time is exceeded or a local optimum is found. Computing all disjoint insertion costs for each unconnected cabinet each time, leads to the best results, but is too slow to remain within a computation time of at most several minutes for large instances of up to 10,000 vertices, 20,000 edges and 600 cabinets. Therefore, non-disjoint insertion costs are used to determine the insertion order. Furthermore, the influence of the initial

2.4 Summary

number of rings was investigated. The analysis showed that if little time is available for the local search, it is better to choose extra rings for the initial solution. If much time is available for the local search, however, it is wiser to choose no or only few extra rings. After tuning the different options in the Insertion Heuristic, it was tested against the CFRS heuristic. The Insertion Heuristic showed very good performance compared to the CFRS Heuristic. Only for a few tested instances, the CFRS Heuristic showed lower average cost (1% to 3%) over 100 replications: medium-sized instances with no or very few inexpensive ducts. For small and large instances with no or few inexpensive ducts and instances with more inexpensive ducts, the Insertion Heuristic gave solutions that were on average between 1% and 27% better. The computation time of the Insertion Heuristic is in most cases larger, but remains well within the set limits. A big advantage of the Insertion Heuristic is that it always finds a solution, whereas the CFRS Heuristic sometimes fails to give a solution especially when the number of cabinets is a bit higher compared to the number of vertices in the graph.

There are still many possibilities for improvement and further research. Firstly, the methods can be tested on real grid instances. In practice obstacles (e.g. canals, railways) can cause having only few possibilities to go in a particular direction. This could have a significant effect on the performance of the methods. The different methods themselves also provide opportunities for further study. For applications where computation time is less important, it is expected that the initial solution quality can still be improved by using disjoint insertion cost information for all cabinets for each ring to decide on the insertion order. Sophisticated objectives that decide which cabinet to insert first, could be used instead of just inserting the cheapest cabinet. In this way, relatively expensive insertions at the end of the algorithm can be prevented a lot better. Another option is to increase the insertion cost artificially as the rings become fuller. In this way, as a ring becomes fuller only cabinets that have no relatively inexpensive insertion alternatives will be inserted. This could prevent that the last insertions are very expensive because the nearest rings are already fully used. To get the right parameter values for the artificial cost increase, extensive testing will be required though. Next to this, the computation time of the local search is quite long. Future research can try to speed it up by looking further into the order of the cabinets that is considered for exchange or relocation. Additionally, maintaining a list of recently unsuccessful relocations and exchanges to temporarily skip them, can probably increase the convergence speed. Finally, it is interesting to see the result of combining different aspects of the CFRS Heuristic and the Insertion Heuristic.

3 FttH planning and economic impact

Fourth generation broadband (4GBB) refers to future service packages that are so demanding in bit rate that they easily consume a bandwidth of hundreds of Mbit/s. What kind of services will be involved in 4GBB is currently unknown, but it will probably include many high definition video channels simultaneously. To deliver 4GBB to end users, a next generation access technology is needed, and the way such technology is implemented is irrelevant as long as it can comply with the high bandwidth demands from 4GBB service packages. The use of fibre will be inevitable for transporting hundreds of Mbit/s to and from end-users, but this does not necessarily mean that fibre is to be deployed all the way to a point into the home, Full Fibre to the Home (Full FttH). An alternative is bringing fibre up or near to the home and reusing existing copper cables, Fibre to the Curb or Fibre to the Building. In this chapter a number of challenges are faced, regarding the introduction of Fibre to the Curb and Full FttH. First planning issues when introducing both network configurations are studied. Next the economic impact is studied in more detail, where not only the investment is considered, but also the possibly missed revenues. To illustrate this, two cases are introduced: what are the cost effects of a total migration path and what are the total economic effects of introducing Hybrid FttH? This chapter is based on [118, 119, 120, 123, 142].

3.1 Introduction

In this section some technical background on the various (hybrid) fibre networks is presented and the problem is placed in the literature.

3.1.1 Hybrid FttH

When it is too expensive or operationally difficult to bring fibre all the way to the end user the current copper cable can be reused for the last part to the customer. However, it is technologically challenging to realize the high bandwidth over this copper cable. The copper technology that is required for such a Hybrid FttH solution is currently developed and is named G.Fast. First results of this development make it plausible that Hybrid FttH using G.Fast is *technically feasible up to 1 Gbit/s*. For details on this work and more information on G.Fast technology we refer to the 4GBB project [23] and Van den Brink [18, 19].

The implementation of Hybrid FttH, using G.Fast, instead of offering Full FttH directly, may seem a very odd idea at the first glance. Why leave the remaining 100-200 meter unchanged if the constructor is already there for construction work? However, reusing existing telephony wiring in streets, apartment buildings and multi-tenant houses could save a lot of digging where installation costs are very expensive especially those last meters. And it may also save much installation time such that operators can quickly respond to a sudden increase in demand for 4GBB. If both FttH solutions are technically available, it might be possible to make an implementation choice on a case by case basis. For instance, when FttH is installed in city areas, then one may decide to offer Full FttH to low-rise buildings and to offer Hybrid FttH to those locations in a street where high-rise buildings, multi-tenant houses or apartment buildings are situated. In such a case, fibre is brought up to a manipulation point in the basement of a building, or to a wall-mount cabinet on the ground floor of an apartment building or even to a foot-way box near the front door of such a group of houses. Next, the fibre optic signal is converted to (and from) electrical signals via a DSL-alike technology as G.Fast. And when a power outlet is lacking to feed these electro-optical converters, then it can be powered from the customer side via the existing telephony cabling. An operator can bring 4GBB connection to a certain area in the most economical way, considering the number of apartment buildings, the building density, the available number of copper pairs etcetera.

The simple fact that we believe that Hybrid FttH is technically feasible and that it might be a bright idea in certain cases, does not mean that it is also feasible from a purely techno-economic point of view. On one hand Hybrid FttH saves cost for installation and for digging into the ground and it may also be quicker to install, but on the other hand it may be more costly in equipment and operational costs. The operator has to introduce a huge number of new active points in the network and has to find out where to place, install and maintain them. To find the right place for the new equipment a very large number of possible locations have to be evaluated, where these locations should comply to a number of requirements. An approach for this is presented in the next chapter. Next to this, the question arises whether the used infrastructure will be reusable for possible migration to Full FttH when then bandwidth seems not to be sufficient in the future. Then double investments have been made. This will be discussed in Section 3.4.

3.1.2 FttH topologies

If a region has made the decision to directly implement Fibre to the Home (FttH), there are still many decisions which still has to be made:

1. Which technology/topology is most suitable?

2. Where will the concentration points be located (or PoP, Point of Presence)?

3. What is the dimensioning of these concentration points?

4. How will the fibres run physically?

The last three questions will be discussed later, here the first question is considered briefly. FttH has three known families of connective networks, namely a passive star (a

3.1 Introduction

Point-MultiPoint topology), home run (a Point-to-Point topology) and active star (also a Point-to-Point topology). In view of the high costs of implementing fibre optic, the right choice of topology is essential:

1. With a passive star infrastructure, a separate fibre has not been laid for every customer, but the capacity of a fibre is shared by all households within the segment to which the fibre is connected. The signal is continuously split up via passive optic splitters. A solution like that offers major cost advantages. The concept of shared use could prove to be a drawback in future capacity expansion.

2. In the case of a home run architecture, individual fibres are laid directly from a connection point with the backbone network to a customer location. The advantage of this approach is the extreme overdimensioning in capacity (every residence its own fibre). An important drawback are the installation costs.

3. With an active star infrastructure, several local stations are connected with the connection point of the backbone network via a fibre optic circuit with extremely high capacity (e.g. 10 Gbits/s). These local stations contain active hardware. In many cases, xWDM (Wavelength-Division Multiplexing) technology is used to increase the capacity of this circuit. Individual fibers are laid from the local stations to the customer location. The advantage of this solution is the use of multiplexing, which reduces the need to synchronize the capacity of the circuit structure to that of the sum of the capacity of the individual fibres in the sub-loop. Often times the circuit structures are prepared for higher band widths, by installing more fibres than currently necessary. In future, the capacity of the shared circuit can be increased by utilizing more of these fibres.

Oftentimes Ethernet technology is used with Point-to-Point topologies like active star and Home Run, while 'Passive Optical Network (PON)' technology is typically used with a PMP topology. In The Netherlands, the active star topology is preferred.

3.1.3 Literature review

As stated before, FttH has three known families of connective networks, namely a passive star, home run and active star. With regard to planning issues for (Hybrid) FttH networks we see different areas touched mostly for PON networks. Not much research can be found on active star topologies. However, some problems are the same in both networks.

Chardy et al. [25] solve the problem of locating splitters and routing fibres within an existing network infrastructure to which a graph is associated with given capacities on the edges. They then perform a graph reduction and a branch and bound algorithm on the resulting mixed integer linear program using CPLEX. This results in calculation times up to one hour.

Bley et al. [15] are designing a two-level FTTx network and recognize the need for small calculation times. They present two Lagrangian decomposition approaches that

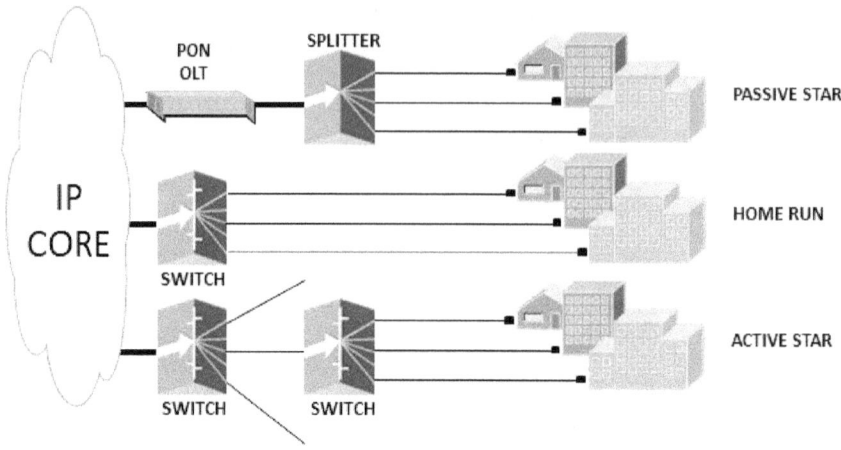

Figure 3.1: PON versus PTP architecture.

decompose the problem based on the network structure and on the cost structure, respectively. The sub problems are solved using MIP techniques. The resulting calculation times are also around one hour.

Mitscenkov et al. [109] address broadband PON access network design which minimizes deployment costs using a heuristic solution. The questions here are: (1) How to form groups of customers that share a PON splitter? (2) Where is the splitter placed? (3) What is the best path from the customer to their splitter unit? (4) How to connect splitters to the central office? The first problem is regarded as a clustering problem which is solved heuristically by combining nearby shortest paths from the customer to the CO. The second problem is simply solved by calculating the optimal location from the (small) set of possibilities. The last two problems are solved by a Steiner Tree problem using a 2-approximation heuristic, the Distance Network Heuristic as presented in Kou et al. [86]. They report calculation times from 10 minutes up to 2 hours.

Li and Shen [95] consider the place of the optical network units (ONU) in a PON and their connection with the CO. They consider two heuristics in their study. The first one is an intuitive sectoring algorithm to divide the area into smaller pieces. The second algorithm, named Recursive Allocation and Location Algorithm (RALA), is an extention of Cooper's algorithm [29] which has been used to solve the Multi-Facilities Location Problem in logistics studies. The RALA algorithm finds an efficient set of splitters as well as the connection relationship between each ONU and the splitters.

Lv and Chen [99] are optimizing the multi-hierarchy planning and fibre routing problem when deploying a PON. The solution to the optimization problem includes the locations and type selections of the Optical Branching Devices (OBD), the hierarchy planning of the PON networks and the fibre routing design. The whole optimization problem is decomposed into two sub-problems; (i) Hierarchy planning, which selects the proper OBDs and plans their hierarchy levels, and (ii) Premium fibre routing algorithm design. Both problems are solved with heuristic based algorithms and the second problem is solved by a genetic algorithm. Almost the same problem is solved by Eira et al. [42] using an ILP model, which is solved for bigger instances using a two stage heuristic.

Ouali et al. [113] consider the problem how to divide the central office area into sub-areas to be planned individually. This paper proposes a tool based on a Mixed Integer Linear Programming (MILP) method that can optimally decide sub-areas while still satisfying given constraints. Each sub-area is served by one fiber cabinet containing the optical splitters. Depending on the distances from customers to the central office and the splitter types being used, the model is capable of identifying customers violating the power budget and assigning them to a splitter with a smaller splitting ratio and longer reach.

Kokangul and Ari [82] solve a multi-hierarchy PON planning problem with system's attenuation and whole equipment's constraints without dividing the problem into sub-problems. A constrained nonlinear integer model, which is intractable due to its non-linearity and NP-hard feature, has been developed for the location allocation problem. To solve this problem, Genetic Algorithm has been developed. To get an applicable result, and find a solution closer to global minimum, local search algorithm and elitism operators are added into a Genetic Algorithm Model.

Considering the techno-economical assessment of access network topologies, the copper technology for Hybrid FttH, more specific G.Fast, is not available yet in these kind of evaluations in the literature, although the migration or choices to be made in the other scenarios have been studied by many projects. However, in most studies the choices are made by the operators, even the choice to wait. We give a major role to the strong demand of the customers and the role the situation of the competition plays in this field. The European projects IST-TONIC [68] and CELTIC-ECOSYS [24] resulted in various upgrade or deployment scenarios for both fixed and wireless telecommunication networks, published in [129] and [110]. A major question in these studies is when to make the decision to roll out a FttCurb/VDSL network or a Full FttH network. Based on demand forecasts, it was shown that it is profitable to start in dense urban areas, wait for five years and then decide to expand it to the urban areas. With the use of real option valuation the effect of waiting is rewarded to identify the optimal decision over time.

In [149] the OASE approaches are presented for more in depth analysis of the FttH total cost of ownership and for comparing different possible business models both qualitatively and quantitatively. OASE stands for Optical Access Seamless Evolution and was performed within the EU FP7 framework.

Casier [22] presents the techno-economic aspects of a fibre to the home network deployment. First he studies all aspects of a semi-urban roll-out in terms of dimensioning and a cost estimation models. Next, the effects of competition are introduced into the analysis.

Antunes et al. [7] present a multi-criteria model aimed at studying the evolution scenarios to deploy new supporting technologies in the access network to deliver broadband services to individuals and small enterprises. This model is based on a state transition diagram, whose nodes characterise a subscriber line in terms of service offerings and supporting technologies. This model was extended for studying the evolution towards broadband services and create the optimal path for broadband network migration. A same kind of model is presented by Zhao in [156], where also an optimal strategy is proposed using a dynamic migration model.

In all those papers G.Fast is not taken yet into consideration. Next to this, we think that incumbent telecom operators need all the effort to keep in track of the cable operators. There is almost no time for sophisticated strategies; they have to connect as much as possible of their clients with a sufficient high bandwidth connection.

3.2 Planning of FttCurb

In this section a simple framework for planning options when deploying FttCurb, using G.Fast as technology, is presented. The general idea here is a framework of eight possible planning options, roll-out scenarios, coming from three main planning choices. The mathematical approach of each of these eight options is elaborated, using combinations of existing methods. Also, the results of a real life case, rolling out FttCurb in Amsterdam and The Hague, is presented, resulting in an example of the calculation time needed and an indication of the costs of such a roll-out.

3.2.1 Background

Here the planning of the Hybrid FttH variant, where the fibre is brought to a place in the street, FttCurb is studied. To realize FttCurb using G.Fast a next step in bringing fibre to the houses is needed. A new node is realized within 200 meter of each house connected. This 200 meter is the assumed maximum distance that G.Fast brings value. Assumed here is that a branching point in the existing copper connections is chosen to place the new active equipment. Technical issues like modulation and power supply are considered in other work of the CELTIC/4GBB project [23]. The new nodes have to be connected by a fibre connection. In this chapter is argued that there have to be made three main choices before designing the network. If these three choices all have two options, there result eight possible roll-out scenarios that are all elaborated in this section.

In the remainder of this section the starting position of the copper network is presented and the main choices that have to be made by the designer of the network are shown. Next the various combinations of those choices are elaborated and the literature for the mathematical approach for those combinations is explored. At the end a real case from two cities in the Netherlands, Amsterdam and The Hague, is discussed.

3.2.2 Identifying the options

In this section the framework based on three questions is presented and the eight planning options that result from these questions are elaborated. Next the choice between a tree and a ring based network structure is discussed shortly .

3.2 Planning of FttCurb

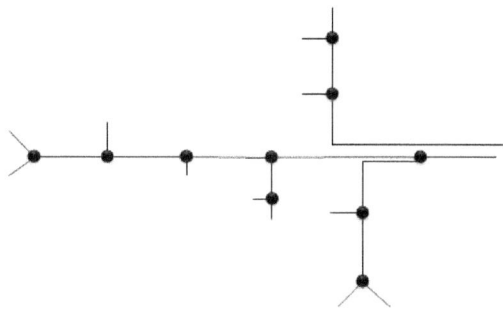

Figure 3.2: Typical last mile in the Netherlands.

3.2.2.1 Three questions

When we look in more detail to this next part of the copper network we see a situation as shown in Figure 3.2. This is a typical situation in the last mile of the Dutch copper network: a heavily branched network, with at the right side a cabinet. In this network new network nodes have to be placed for the G.Fast technology. To do this, possible locations for these network nodes have to be determined, logical places are the dots in the figure, the branching points of the network. Assumed is that it is known which houses are connected to these locations at which distance. Now, one should decide which locations will be used and how they are connected to a fibre node in the most economic way.

The three questions that form the basis of the framework are:

1. Should *all* houses be reached from a Hybrid Fibre node within a fixed distance, or a fixed *percentage* of houses, or is a fine considered for each house that is not connected within that certain distance? Distinguish:

    (a) All houses must be connected, a fine is considered otherwise.

    (b) A certain percentage has to be within the defined distance.

2. Does the node have a capacity restriction?

    (a) Yes.

    (b) No.

3. How are the nodes connected?

    (a) Tree or star structure.

    (b) Ring structure.

3.2.2.2 Eight planning options

In the previous section there were three choices to be made, each having two possible answers. This leads to eight possible roll-out scenarios that are in theory all thinkable. In this section all scenarios are discussed and a mathematical approach to each planning problem is proposed. Each scenario is referred to with a three letter acronym, each representing the chosen answers to the questions. As example, the scenario AAA refers to the case where all questions were answered by option (a): all houses connected, the node has a capacity restriction and the nodes are connected by a tree or star structure. The eight possible roll out scenarios are then:

- AAA (CFLP plus MSTP): The scenario AAA refers to the case where all houses have to be connected, the node has a capacity restriction and the nodes are connected by a tree structure. This problem can be seen as the case where from several possible facilities with a certain maximum capacity a subset of those facilities has to be chosen and customers have to be assigned to a facility such that all customers are served by one facility at minimal cost. This is a Capacitated Facility Location Problem (CFLP). Next the opened facilities have to be connected with the central point (cabinet, CO) in a star structure. To do this the shortest path between the central point and the opened facilities can be determined, but to reduce the cost of digging it is more economical to take the minimal spanning tree between all the facilities and the central point by solving a Minimal Spanning Tree Problem (MSTP).

- AAB (CFLP plus VRP): The scenario AAB refers to the case where all houses are connected, the node has a capacity restriction and the nodes are connected by a (multiple) ring structure. This looks like the previous problem, only now the routing comes into scope. The central point uses ring structures to serve the opened nodes in a shortest cycle. Which ring has to serve which node and what is the shortest path the ring has to go? This is a Vehicle Routing Problem (VRP), or if there is a maximum number of nodes that can be connected in one ring a Capacitated Vehicle Routing Problem (CVRP).

- ABA (standard FLP plus MSTP): The scenario ABA refers to the case where all houses are connected, the node does not have a capacity restriction and the nodes are connected by a tree structure. This is a standard or uncapacitated Facility Location Problem (FLP). Again the Minimum Spanning Tree Problem can be used to connect te opened facilities.

- ABB (standard FLP plus VRP): The scenario ABB refers to the case where all houses are connected, the node does not have a capacity restriction and the nodes are connected by a ring structure. This is an uncapacitated Facility Location Problem in combination with a Vehicle Routing Problem.

- BAA (activation problem plus MSTP): The scenario BAA refers to the case where a certain percentage of the houses have to be within the defined distance, the node has a capacity restriction and the nodes are connected by a tree structure.

3.2 Planning of FttCurb

This is the same problem as discussed in Section 2.2.1 for VDSL cabinet activation, combined with the Minimum Spanning Tree Problem to connect the opened facilities.

- BAB (activation problem plus VRP): The scenario BAB refers to the case where a certain percentage of the houses have to be within the defined distance, the node has a capacity restriction and the nodes are connected by a ring structure. This is again the activation problem, combined with the Capacitated Vehicle Routing Problem (CVRP) to connect the opened facilities.

- BBA (activation problem plus MSTP): The scenario BBA refers to the case where a certain percentage of the houses have to be within the defined distance, the node does not have a capacity restriction and the nodes are connected by a tree structure. This is the activation problem, now with infinite capacity on the nodes. Again combined with the Minimum Spanning Tree Problem to connect the opened facilities.

- BBB (activation problem plus VRP): The scenario BBA refers to the case where a certain percentage of the houses have to be within the defined distance, the node does not have a capacity restriction and the nodes are connected by a ring structure. This is the activation problem, now with infinite capacity on the nodes combined with the Capacitated Vehicle Routing Problem (CVRP) to connect the opened facilities.

If we look at these eight roll-out scenarios and the identified standard problems, these can be summarized in the following six problems:

1. scenario AAX: Capacitated Facility Location Problem (CFLP).
2. scenario ABX: Uncapacitated Facility Location Problem (FLP).
3. scenario BAX: Activation Problem.
4. scenario BBX: Activation Problem with infinite node capacity.
5. scenario XXA: Minimum Spanning Tree Problem (MSTP).
6. scenario XXB: (Capacitated) Vehicle Routing Problem (CVRP).

The total framework can now be summarized in a flow diagram, as depicted in Figure 3.3.

3.2.2.3 Ring or star

One of the choices to be made was the choice between a star or tree and ring topology. In the Netherlands ring structures are common, but in other European countries star or tree topologies are conventional. Mostly cost are the driver for this choice. Ring topology deliver a much higher reliability however, and the break-even costs (in terms of distance of digging) where both topologies are equally expensive is reached fairly rapidly (ring vs star) or are close all the time (ring vs tree).

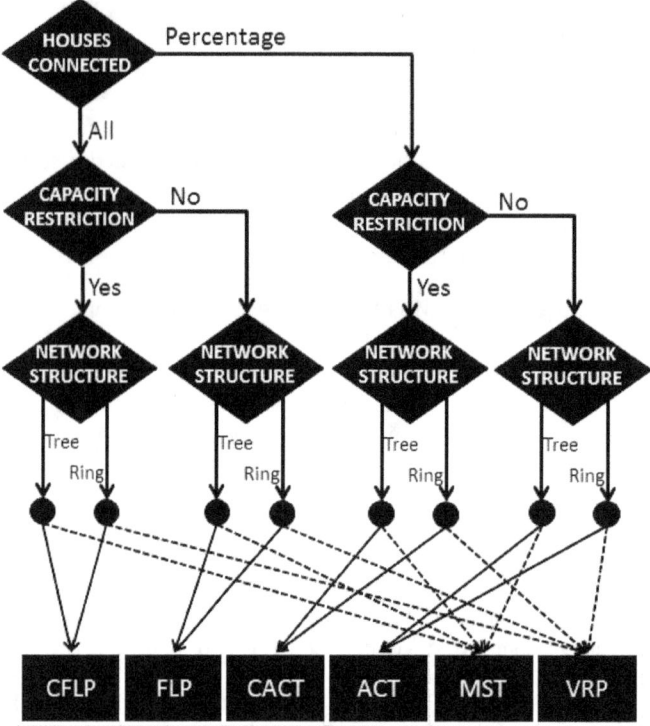

Figure 3.3: Framework flow diagram.

Theoretically the break-even point between the star and the ring structure can be derived very simply. Say we have n nodes, all of them at the distance r of a centre point. Connecting them with a star structure will cost $n \cdot r$. If a ring structure is chosen, all nodes are on the ring with radius r, the total costs are

$$\frac{n-1}{n} \cdot 2 \cdot \pi \cdot r + 2 \cdot r.$$

The break even point is where

$$n \cdot r = \frac{n-1}{n} \cdot 2 \cdot \pi \cdot r + 2 \cdot r.$$

This is true when

$$n = \pi + 1 + \sqrt{\pi^2 + 1} = 7.44.$$

This means that if eight or more nodes have to be connected a ring structure is cheaper.

However, it is obvious that the nodes will not be distributed such that they are all at distance r from the CO. To find the relation between the number of nodes and the digging length of the three network structure options 1000 situations have been simulated, where

3.2 Planning of FttCurb

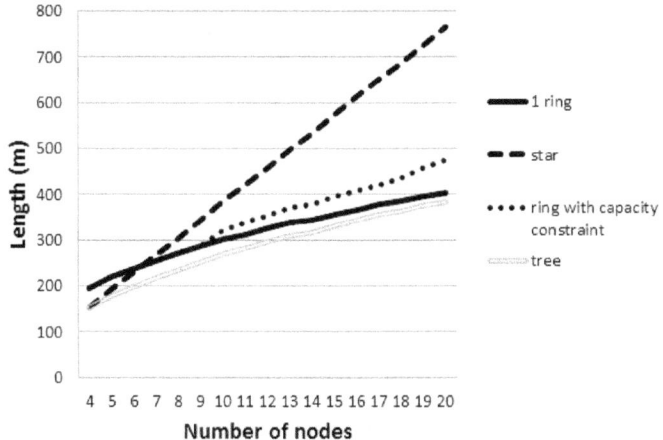

Figure 3.4: Digging distances as a function of the number of nodes.

n nodes are placed randomly within an area with dimension 100×100. The central node is placed at $(x, y) = (50, 50)$. For each situation the n points are connected with the central node in a star structure, in a tree structure, in a ring structure and in a ring structure with at most 10 nodes per ring. The ring is created solving a TSP using a generic insertion algorithm and 2-opt algorithm. The ring with capacity constraint is created solving a CVRP using Clark and Wright savings algorithm [91]. The tree is calculated using Prim's algorithm [124]. The results are shown in Figure 3.4. Here we see that the break even point for star vs ring is between six and seven nodes and the that the tree is always slightly cheaper than the ring structures. The owner of the network has to weight this against the differences in reliability of the structures.

3.2.3 Standard problems

In the previous section six problems were identified that appear when affect the eight roll-out scenarios. In this section for each problem an overview of literature dealing with that problem is given.

3.2.3.1 Uncapacitated facility location problem

The FLP revolves around the following problem: given a set F of facilities, a set D of customers, costs f_j for opening facility $j \in F$ and connection costs c_{ij} for connecting the customer i with facility j: which subset of facilities in F needs to opened and which customers have to be connected with which open facility, in order to minimize the costs. A facility is considered open when at least 1 customer is serviced by this facility. Opening a facility and servicing customers involves costs. To coop with the maximum distance of 200 meters one should express that in the cost parameter c_{ij}.

Literature (e.g., [102]) makes a distinction between several different types of FLP. The difference is important, as the known heuristics used cannot be the same for all types.

The first important type here is the Uncapacitated Facility Location Problem (UFLP). The assumption there is that the capacity of a facility, or the number of customers which can be serviced, is infinite, and the costs of opening a facility are set. So the opening costs of a facility are not determined by the number of customers serviced.

Again in [102] a metric Uncapacitated Facility Location Problem (UFLP) is discussed. The connection cost is metric as they are symmetrical and meet the triangle inequality. The article discusses first the JMS heuristics. The heuristic is presented in Algorithm 8. Next they present a more complex, but also more efficient algorithm.

Algorithm 8 JMS based on [102].

Require: A start solution start: all customers are unconnected and all facilities closed. The budget of every customer i, noted with B_i is equal to 0.
Ensure: An assignment of all customers to a set of opened facilities.
1: **repeat**
2:   Increase the budget of each unconnected customer by the same value ϵ.
3:   If for an unopened facility j, the total offer which facility j receives from all customers is equal to the costs of opening facility , then facility j is opened and for each customer i (serviced or not serviced) that has an offer to facility j greater than 0, this customer i is connected with facility j.
4:   If for a non-serviced customer i and an already opened facility j the budget of customer i equals the connection costs c_{ij}, then this customer i is connected with facility j.
5: **until** All customers are assigned.

3.2.3.2 Capacitated facility location problem

Where the UFLP can be solved relatively easy by a good and simple heuristics, adding capacity constraints to the facilities makes the problem much more difficult. Most research on the CFLP has focused on the development of efficient solution algorithms, based on branch-and-bound techniques, Lagrangian relaxation, Benders decomposition etcetera, see for example [64], [107] and [134]. Gollowitzer [56] defines the Capacitated Connected Facility Location Problem (CapConFL) for a similar problem. A nice local search heuristic can be found in [9]. An other possibility is to use the solution to the Activation Problem of the next section, with 100% customers connected.

3.2.3.3 Activation problem

In Section 2.2.1 this problem is discussed for the FttCab roll-out. This works for both the activation problem with and without infinite node capacity. The problem there is: which cabinets must be activated in order to reach the desired percentage of households at minimal costs? All cabinets are connected through copper with the CO. Several residences are connected to the cabinet. Now a subset of the cabinets needs to be activated in order to reach the intended number of households over copper from an activated cabinet within the set distance. In fact, this is a generalization of the CapConFL.

The proposed heuristic starts with a logical, allowed, solution, in which all cabinets are activated in step 1. Next in step 2, all possible cascade arrangements are determined and the savings of this arrangement (call it B) as well as the number of customers which as a result are positioned outside the desired distance of, here, 200 meter (call it K)

3.2 Planning of FttCurb

are reviewed. Next, the solutions which generate a saving ($B > 0$) can be sorted by two possible characteristic: B and B/K. In step 3 that solutions are realized that have the largest (negative) value of B or B/K, until the requirement of, e.g., 90% of the customers is reached. In step 4 we perform a 2-opt approach to improve the solution. If step 4 results in a swap, a new improvement has to be found; if no swap could be found the best solution was found. This algorithm is very fast as shown earlier in this thesis.

3.2.3.4 Minimum spanning tree problem

Given a connected, undirected graph, a spanning tree of that graph is a connected sub graph, connecting all the vertices of the original graph. If the edges have a weight assigned, these weight can be used to compute the weight of the spanning tree, the sum of the weights of the edges in that spanning tree. A minimum (weight) spanning tree is then a spanning tree with weight less than or equal to the weight of every other spanning tree. A solution to MSTP can be found in [124]. An alternative is the method of Kruskal. A nice comparison can be found in [58]. Prim's algorithm is quite simple:

1. Take some arbitrary start node s. Initialize tree $T = \{s\}$.

2. Add the cheapest edge, which has one vertex in T and one vertex not in T, to T.

3. If T spans all the nodes the Minimum Spanning Tree is ready, else repeat step 2.

3.2.3.5 Capacitated Vehicle routing problem

The Vehicle Routing problem comes from logistics and describes the problem that clients have to be serviced from (one or more) depots, using one or more vehicles that might have a certain capacity constraint. The question in this problem is which client is serviced by which vehicle from which depot and what is the shortest route the vehicle will drive. Two main questions in our problem will be: which node is serviced by which ring and how does the ring run physically. To solve these two problems together the best-known approach is the 'savings' algorithm of Clarke and Wright. Its basic idea is very simple, as described in [91]: 'Consider a depot D and n demand points. Suppose that initially the solution to the VRP consists of using n vehicles and dispatching one vehicle to each one of the n demand points. The total tour length of this solution is, obviously, $2\sum_{i=1}^{n} d(D,i)$. If a single vehicle is used to serve two points, say i and j, on a single trip, the total distance travelled is reduced by the amount

$$\begin{aligned} s(i,j) &= 2d(D,i) + 2d(D,j) - [d(D,i) + d(i,j) + d(D,j)] \\ &= d(D,i) + d(D,j) - d(i,j). \end{aligned}$$

The quantity $s(i,j)$ is known as the 'savings' resulting from combining points i and j into a single tour. The larger $s(i,j)$ is, the more desirable it becomes to combine i and j in a single tour. However, i and j cannot be combined if in doing so the resulting tour violates one or more of the constraints of the VRP.' Where $d(i,j)$ is the distance function.

The algorithm can now be described as follows.

1. Calculate the savings $s(i,j) = d(D,i) + d(D,j) - d(i,j)$ for every pair (i,j) of demand points.

2. Rank the savings $s(i,j)$ and list them in descending order of magnitude. This creates the savings list. Process the savings list beginning with the first item.

3. For the savings $s(i,j)$ under consideration, include link (i,j) in a route if no route constraints (e.g. the capacity of the vehicles) will be violated through the inclusion of (i,j) in a route, and if:

    (a) Either, neither i nor j have already been assigned to a route, in which case a new route is initiated including both i and j.

    (b) Or, exactly one of the two points (i or j) has already been included in an existing route and that point is not interior to that route (a point is interior to a route if it is not adjacent to the depot D in the order of traversal of points), in which case the link (i, j) is added to that same route.

    (c) Or, both i and j have already been included in two different existing routes and neither point is interior to its route, in which case the two routes are merged.

4. If the savings list $s(i,j)$ has not been exhausted, return to Step 3, processing the next entry in the list; otherwise, stop: the solution to the VRP consists of the routes created during Step 3.

However, to fully exploit the reliability gain of a ring structure, all the elements (paths) of the ring should be independent. The ring should not use the same trench or cable twice (or more). This is not taken into account in a regular CVRP solutions, like Clarke and Wright [28].

Kalsch et al. [75] developed a mathematical model and a heuristic approach for embedding a ring structure in a fibre network, that takes into account the following restrictions: ensuring a ring structure, a maximum number of nodes in a ring, each node in exactly one ring, and that the ring uses each edge only once. It is, however, hard to draw conclusions on the performance of their approach, since no further information is given on the data used for a test case. Another important disadvantage of their method is that no real attention is paid to the clustering of the nodes to the rings. They indicate clustering is part of the problem, but do not really treat it in there article and they go directly to the routing part of the problem.

A similar problem in the designing of a FttCab network the two problems, clustering and routing, has been solved by us in succession see Sections 2.2.2 and 2.2.3 and together in Section 2.3.

3.2.4 Cases

Here scenario BBA is performed to two cities in the Netherlands, Amsterdam and The Hague, using the activation algorithm of Section 2.2.1 and Prim's algorithm [124]. Assumed is that the cabinets already have a fibre connection, so our focus is the part of

3.2 Planning of FttCurb

Description	Costs
G.Fast multiplexer	€ 25 per port
G.Fast Manhole	€ 500
Digging and cables	€ 25 per meter

Table 3.1: Costs input.

the network between the cabinet and the home connection. The Amsterdam case has 150,058 branching points in that area, The Hague has 89,076 branching points. Those branching points are the potential spots to place the new equipment. In Figure 3.5 an example is shown: a part of Amsterdam with all the splices and cabinets. The pictures comes from the Giant/PlanXS tool of TNO, which performs FttCab and FttCurb planning problems. In both cities we want to connect at least 99% of the customers within 200 meter to a G.Fast node. Each G.Fast node is placed in a manhole. Each combination of 16-port and 48-port G.Fast equipment (G.Fast multiplexer) can be placed in the manhole.

The problem solved by the activation algorithm is the following: which nodes should be activated in order to reach the desired percentage of households at minimal costs? A household is reached when the distance over copper is less than a chosen length, here 200 meter. Households which meet this requirement are said to meet the distance requirement. This can be solved using the model and heuristic of Section 2.2.1.

However, there are two specific important constraints that were not in that specific model:

1. One arriving cable at the G.Fast node cannot be spread over 2 G.Fast multiplexers.

2. Maximum distance over copper to the active point.

For the first constraint look at the example in Figure 3.6, two cables arrive at node A, one with 14 connections and one cable with 9 connections. If the capacity of the multiplexer is 16, node A can be used to handle both cables with two multiplexers. However, these $14 + 9 = 23$ cables arrive at node B in one cable. This cable cannot be handled with one multiplexer, thus these cables should be handled by an activated node before node B. In the activation problem (see Section 2.2.1) this situation should be depicted in the parameter b_{ij}, meaning: handling the connections of node i by node j keeps b_{ij} connections within the desired distance. However the distance from the home connections to B or C might be less than the chosen maximum copper length 200 meter, we have to make b_{ij} with $i = A$ or lower in the network and $j = B$ or higher in the network equals zero to prevent handling the connections at B or C. If we have also a 48 port multiplexer, this connection can be handled by node B or C.

Also the length constraint, the second constraint, should be depicted in the parameter b_{ij}. Here, with a node capacity of 48 connections and equipment capacity of 48 connection, $b_{AA} = 23$, $b_{AB} = 23$, $b_{AC} = 23$, but $b_{ACab} = 0$. The length to the cabinet is more than 200 meters. The minimal cost selection of new locations for these G.Fast nodes is then connected by a Minimum Spanning Tree. The assumed costs are shown in Table 3.1.

The calculation time for the case Amsterdam is 50 seconds, consisting of:

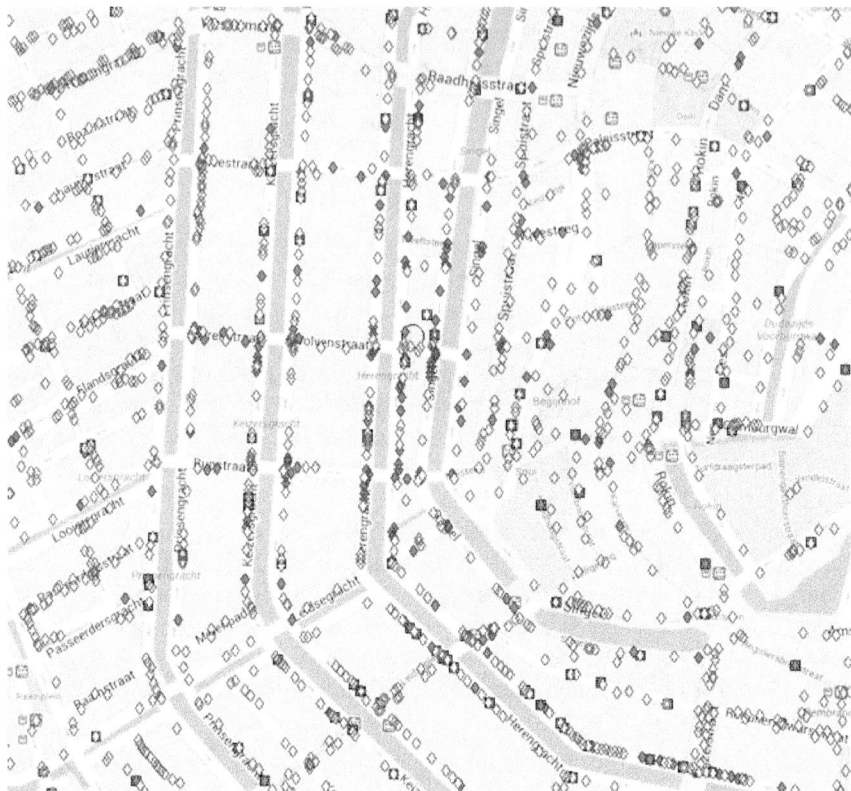

Figure 3.5: Part of Amsterdam showing all the splices (open dots) and cabinets (closed dots).

- 14 spanning tree calculations: 1 second.
- Database interaction and data handling: 25 seconds.
- Solving 2745 Activation Problems: 24 seconds.

The activation problem activates 7,366 new G.Fast nodes, out of the possible 150,058, for 490,000 connections in Amsterdam.[1]. The results of this calculations are presented in Table 3.2 This means we have a port utilization[2] of 74% and an average digging distance per node of 93 meters. For various distances, the costs per home connected is depicted in Figure 3.7. Note that we do not make extra nodes, next to the existing branching points, thus the minimum distance is restricted by the length of the last piece of copper in the path towards the houses. A copper length of 25 meter does not indicate that all copper lengths are lower, but only those connections who can physical realize this. Otherwise the graph is expected to increase faster when decreasing the distance. The

---

[1]There are more ports than connections, due to the fixed number of ports per multiplexer.
[2]Number of connections divided by number of ports.

3.2 Planning of FttCurb

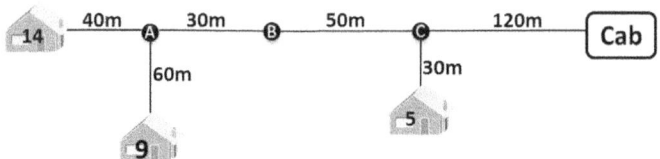

Figure 3.6: Example of cable with connections.

Description	Quantity	Cost (€)
Digging (meter)	686,106	17,152,650
Equipment (ports)	661,024	16,525,600
Manholes (new node)	7,366	3,683,000
Total costs (euro)		37,361,250
Per connection (euro)		76.19

Table 3.2: Results Amsterdam.

trend line indicating this in the figure is an estimation of the real relation, based on a logarithmic trend.

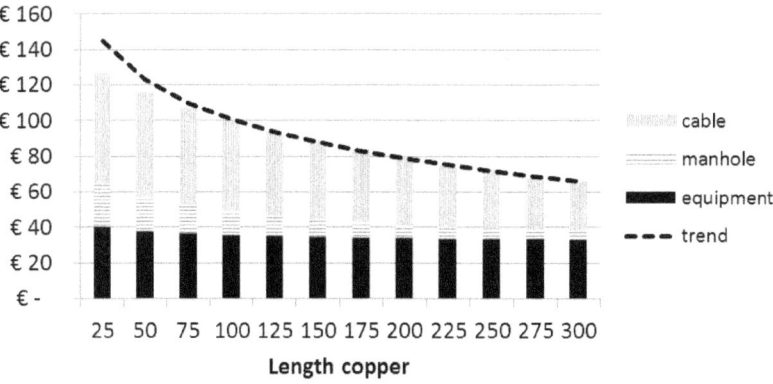

Figure 3.7: Costs of connection in Amsterdam.

For The Hague the results are presented in Table 3.3 and in Figure 3.8. Here we have 288,000 connections, resulting in a port utilization of 73% and an average digging distance per node of 122 meters. The difference between Amsterdam and The Hague are explained by the existing copper infrastructure. Amsterdam-region has already 60% of the homes within 200 meters of the cabinet, and Amsterdam-Centre even 75%. The Hague only has 28% of the connections within 200 meters.

Description	Quantity	Cost (€)
Digging (meter)	1,058,350	26,458,750
Equipment (ports)	395,328	9,883,200
Manholes (new node)	8,656	4,328,000
Total costs (euro)		40,669,950
Per connection (euro)		141.15

Table 3.3: Results The Hague.

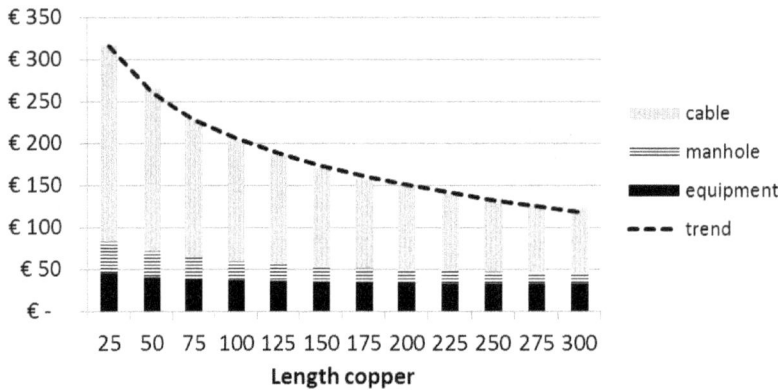

Figure 3.8: Costs of connection in The Hague.

3.3 FttH planning

Rolling out a Fibre to the Home architecture is very expensive, mainly due to all the trenching that is very labour extensive. In this chapter a method is presented for planning the last mile in FttH in *a cost effective way* and extend this method for incorporating and *minimizing the inconvenience and social costs* this work brings. To do this, a model to determine the best place for the concentration unit of a FttH network is presented. This problem can be solved easily by the well-known JMS model. Next some extensions to the base model are presented and shown is how these can be solved. Next the problem of connecting the houses with FttH is modeled in a novel way: where the fibre segments has to be is assumed to be known, but how to connect them to create a coherent network? This can be modelled as a Steiner Tree. Next a heuristic approach is presented to solve this node and edge weighted minimum Steiner Tree Problem and introduce a method to incorporate smart co-laying opportunities in this approach, the timing benefits.

3.3.1 Background

Rolling out a FttH architecture is very expensive, mainly due to all the trenching that is very labour extensive. The main topic of this section are the last three questions of Section 3.1.2. Especially the last seems trivial, but do not forget that a lot of money

3.3 FttH planning

Figure 3.9: Saving costs by digging yourself (source: Trouw 2013).

can be saved here for the owner of a fibre optic network when this is done smartly. This part of the network is (per connection) the most expensive. Digging ditches, which have relatively few connections running through, is both labour and cost intensive. Shortening the digging distances, (re-)using existing ditches and pipes can save a lot of money. Off course you can dig yourself as an example in the town of Bergeijk showed (see Figure 3.9). Next to this, all this labour in your street will cause much inconvenience. How can this been taken into account when planning the FttH roll-out? An other important aspect of our work is the practical application. The algorithm should be easy to implement and fast. Most researchers say that complex problems like the one discussed here are only done once, saying the calculation time is no issue. However, our experience is that planners want a interactive tool; the resulting solution have to be manually adjusted due to the unreliable data and the planners have to plan a great number of areas. For example, even a small country such as the Netherlands has over 6,000 areas to be planned for the FttH roll out.

In this section solutions are presented to solve the location and dimension of the PoP in Section 3.3.2 and a method to optimize a combination of investment and social

Figure 3.10: Difficult access to stores.

costs in Section 3.3.3. Bringing fibre to the premises is seen by many municipalities as an attractive incentive for attracting or keeping residents and companies. However, the installation causes a major inconvenience. Large portions of the public streets must be broken up which cause difficulty accessing stores (see Figure 3.10)[3], detours, temporarily or permanently damaged greenery, and so forth.

This inconvenience can be avoided by planning ahead. When planning the installation of a fibre optic network there are several possible options. First, combining the installation of a fibre optic network with other infrastructural projects like renovating the sewer system, or resurfacing the road can reduce many of these unnecessary inconveniences. This is also known as smart co-laying, which prevents many nuisances and is cheaper as well.

A second possible solution is to replace traditional digging by using trench-less techniques. This may involve drilling during which the pipe is drilled or compressed into the ground without having to break open the street. Generally, drilling is more expensive than digging, however, when it significantly reduces social costs, then it bears consideration from a municipal perspective.

A third possible solution is adapting the network structure at busy traffic intersections or business areas and choosing detours so that this area is not touched. Adapting the structure in this manner results in higher expenses and more meters to trench, but

---

[3] In Bergeijk (The Netherlands) inhabitants dig their own cables to lower the costs of the connection.

will reduce social costs. Generally, the pros and cons of the social costs of a project are not taken into account when analyzing the economic cost-benefits of business opportunity. In order to include such effects in the installation of a fibre optic network, they will have to be expressed in a monetary value. An example of monetizing the inconvenience would be to assign a value to the loss which results from a detour caused by a roadblock. This can be done with the simple algorithmic rule of: the number of vehicles per 24 hours x average detour time x travel time appreciation in Euro equals the monetary value of loss. If we apply this rule to approximately 2,500 vehicles per 24 hours, 10 minutes (= 1/6 hour) detour time and a travel time appreciation of 15 Euro per hour, then we arrive at a loss of $2,500 \cdot (1/6) \cdot 15 = 6,250$ Euro per 24 hours. The same method may be used to determine lost revenue for businesses if they are inaccessible or difficult to access for a certain period of time.

All of these possible options for the installation of a smarter fibre optic network as well as models for calculating the social costs have been combined in a decision support tool. [4] This tool minimizes both the social and direct costs for installing the network. For all of the connection options both costs have been determined and summed with weights applied to the values. The weighting is influenceable. The tool then starts calculating with these summed costs. If social costs are assessed as not important and get a low weight, then the application gives the cheapest solution based on only the direct investments and expenses. A design like that will be predominantly realized by traditional digging. However, if a lot of importance is given to avoiding inconveniences, then the design of the network looks different and alternative techniques for its realization will be offered more often. This creates a unique method for municipality and the intended installation party to decide the best approach together before installation. This also immediately clarifies the costs for avoiding inconveniences. The techniques that are used within the tool are presented in this chapter.

3.3.2 Place of the concentration unit

When implementing FttH, every residence needs to be connected with a concentration point via fibre optic (PoP or Concentration Point, CP). This can be an existing point or a newly created one. The main issue with implementation therefore is: determine the number of CPs and their locations, and allocating households to the CPs. Do this in such a manner that the costs for installing the CPs and laying the fibre optic cables are minimal. This is a question of a Facility Location Problem: Where and which facilities are opened, and which customers are serviced by which facility.

3.3.2.1 Problem definition

To solve this problem the area is divided into multiple areas, so-called pixels. In general the rule is that the smaller the pixel size, the more accurately a solution can be determined. Every pixel has received a number to distinguish them from each other. Each pixel number is assigned a $x-$ and $y-$coordinate to recognize its location. The upper left pixel is the pixel with coordinates $(x,y) = (1,1)$. The coordinates of the other pixels

---

[4]This tool was developed in the TNO project 'Masterglass'.

are relative compared to the upper left pixel. It is permissible to open a CP in every pixel.

In fact it should be determined for each pixel how much fiber optic is needed to connect each household with a CP. The model only determines how much fibre optic is needed to connect the centre point of the pixel with a CP, taking the number of households within the pixel into account. The model assumes that all households are located in the middle of the pixel, or that the households are spread out over the pixel in such a manner that assuming they are in the middle of the pixel has the same effect. Because when the households are spread out extensively over the pixel, then some households need a longer fibre optic cable to connect to a CP, and other households need a shorter cable because they are closer to the CP. On average the length of fibre optic cables will be the same as when the assumption is made that all households are located in the middle of the pixel.

As the model assumes that each CP is in the middle of a pixel in order to calculate fibre optic costs, it can only be determined whether or not a pixel should have a CP. The location of the CP within the pixel is not determined by the applied heuristics.

The total connection cost is determined by the connection costs between the CO and the CPs, and the connection costs between the households and the CPs. The connection costs of the CO with the CPs are initially not taken into account (or set equal to 0). The connection cost is included as follows:

$$c_{ij} = ah_i d_{ij},$$

in which:
c_{ij} = the connection costs for connecting pixel with CP;
h_i = the number of households within pixel;
d_{ij} = the distance between pixel and CP in meters;
a = the costs of fibre optic per meter.

The number of households per pixel and the costs of fibre optic per meter are known. The distance between the pixel and the CP is calculated by using the Manhattan-norm. The Manhattan-norm determines the distance $l(x,y)$ between the pixels $x = (x_1, x_2)$ and $y = (y_1, y_2)$ as follows:

$$l(x,y) = |x_1 - y_1| + |x_2 - y_2|.$$

So $d_{ij} = pl(x^i, x^j)$, where x^i represents the vector of coordinates of pixel i and p is a scaling factor. The scaling factor p is equal to the real length of a pixel (in meters).

The costs per CP are modelled as a constant per pixel which indicates the costs necessary to establish a CP in this pixel. These costs are independent of the number of households in the pixel. These opening costs may differ per pixel.

Now the next model is created: Determine the number and location of CPs to be opened, as well as their locations, and determine for each pixel which CP the households need connecting with, in such a manner that the total costs (all opening and connecting costs together) are minimized. The above model is an 1 (layer)-FLP. It can be formulated as a binary LP-model as follows:

3.3 FttH planning

Define:
$$\begin{aligned} I &: \text{set of pixel numbers;} \\ J \subseteq I &: \text{set of potential locations for CPs;} \\ b_j &: \text{costs for opening a CP on location } j, j \in J; \\ c_{ij} &: \text{connection costs for connecting pixel with CP } i \in I, j \in J. \end{aligned}$$

The decision variables, for $i, j \in J$, are:

$$x_j = \begin{cases} 1 & \text{if CP is opened at location } j, \\ 0 & \text{otherwise.} \end{cases}$$

$$y_{ij} = \begin{cases} 1 & \text{if pixel } i \text{ is connected to CP } j, \\ 0 & \text{otherwise.} \end{cases}$$

Then

$$\min \sum_{j \in J} b_j x_j + \sum_{i \in I} \sum_{j \in J} y_{ij} c_{ij}, \tag{3.1}$$

under the conditions:

$$\sum_{j \in J} y_{ij} = 1 \quad (i \in I), \tag{3.2}$$

$$y_{ij} - x_j \leq 0 \quad (i \in I, j \in J), \tag{3.3}$$

$$x_j, y_{ij} \in \{0, 1\}. \tag{3.4}$$

Condition (3.2) ensures that every pixel is connected to exactly one CP. Condition (3.3) ensures that all pixels will only be connected to an opened CP. Condition (3.4) ensures that only complete CPs are opened and that a pixel is connected to a CP or not.

3.3.2.2 Facility location problem

Ever since approximately 1960, a solution is being sought for the Facility Location Problems (FLP). The FLP was already discussed in Section 3.2. The FLP revolves around the following problem: given a set J of facilities, a set D of customers, costs f_j for opening facility $j \in J$ and connection costs c_{ij} for connecting the customer i with facility j: which subset of facilities in J needs to opened and which customers have to be connected with which open facility, in order to minimize the costs. A facility is considered open when at least 1 customer is serviced by said facility. Opening a facility and servicing customers involves costs. Figure 3.11 shows an example of an FLP.

Here we have four facilities which can be opened, and fourteen customers which need to be serviced by a facility.

A possible solution to this problem is shown in Figure 3.12 shows three opened facilities, and every customer being serviced by a facility.

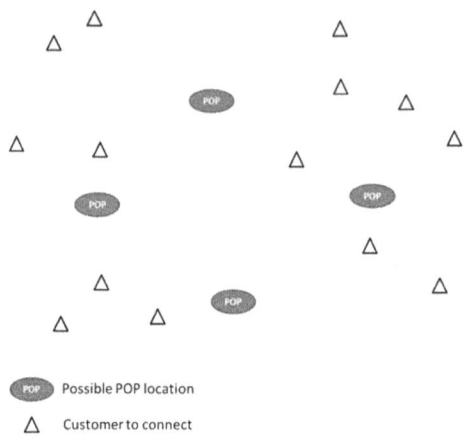

Figure 3.11: Example of an FLP.

Literature (e.g. [102]) makes a distinction between several different types of FLP, which will be discussed briefly here. The difference is important, as the heuristics used cannot be the same for all types.

The first type is the Uncapacitated Facility Location Problem (UFLP). The assumption here is that the capacity of a facility, or the number of customers which can be serviced is infinite, and the costs of opening a facility are set. So the opening costs of a facility are not determined by the number of customers serviced.

A second known type is the Linear Facility Location Problem (LFLP). Here too, it is assumed that the capacity of a facility is limitless, but the opening costs of a facility will increase proportionally to the number of customers serviced. Besides set costs, there are also linear opening costs. In more detail, a LFLP(a, b, c) model is a LFLP model, in which the opening costs O of a facility are structured as follows:

$$O = b + ka,$$

where:

a = marginal costs;
b = set up costs;
c = connection costs per customer;
k = number of customers to be serviced

In principal, the parameters (a, b, c) may differ per facility and may also be different per customer.

A variant of the LFLP is the Soft-Capacitated Facility Location Problem (SCFLP). Here, the openings costs progressively depend on the number of customers. As it happens, a device in the CP has a certain capacity which results in a limited number of households being able to connect with the device. However, when more households are connected with the device in the CP, this can be done by installing a second device in the CP. The opening costs for the CP are then paid twice. In that case the capacity is

3.3 FttH planning

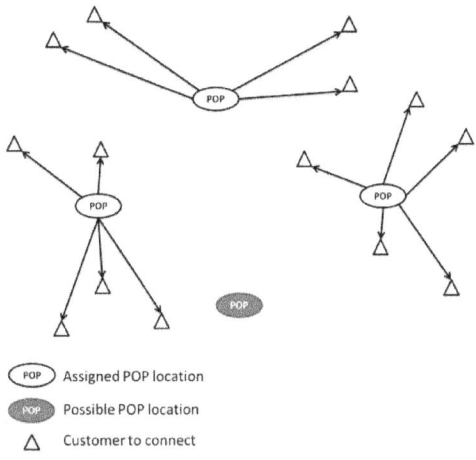

Figure 3.12: Example of an FLP.

not limiting the number of households connected to the CP. The costs of the SCFLP can be formulated as follows, for $j \in J$:

$$O_j = f_j \lceil k_j / u_j \rceil,$$

in which:

O_j : opening costs of one or more devices in CP_j;
f_j : constant costs for opening one device in CP_j;
k_j : number of households connected to CP_j;
u_j : capacity of a device in CP_j.

3.3.2.3 Heuristics

Recall that Mahdian et al. [102] discusses a metric Uncapacitated Facility Location Problem (UFLP) as shown in Section 3.2, Algorithm 8. The connection cost is metric if they are symmetrical and meet the triangle inequality. Next the article discusses the JMS heuristics. The performance ratio is 1.52. This means that in the worst case scenario the JMS heuristics determine a solution in which the costs of the solution are a maximum of 1.52 times higher than the costs of the optimum solution.

For the problem that was described the JMS heuristic can be used. With the implementation of the JMS heuristics, a choice was made to continuously increase the budget by a constant amount ϵ. The budget will be increased until all pixels are connected to a CP. By determining the budget in this manner, the budget assumes discrete values. This results in the necessity to define events in the following manner:

1. If the total offer for an unopened CP j is more than or equal to the costs of opening CP j, then CP j is opened, and for every pixel i (connected or unconnected) which has an offer to CP j more than 0, pixel i is connected to CP j.

2. If, in the case of an unconnected pixel i and an already opened CP j, the budget of pixel i is more or equal to the connecting costs c_{ij}, then pixel i is connected to CP j.

In general, the smaller the value of ϵ, the more precise the solution, but this is not true for all cases. If ϵ is smaller it takes more steps to determine a solution, which extends the calculation time. An advantage of this discrete increase of the budget is that the implementation of the JMS heuristics is easier.

3.3.2.4 Improvements of the base model

In the problem description some assumptions have been made to keep the model simple. However, these assumptions are somewhere a restriction in the quality of the results of the heuristics. In this section we focus on some of these restrictions and propose improvements.

3.3.2.4.1 Connection costs CP to CO

In the base model we did not look at the costs to connect the CP to the existing network, i.e. the CO location. In a star topology these costs are known for each CP:

$$f_{j,CO} = d_{j,CO} e,$$

where:

$f_{j,CO}$ = the connection costs to connect CP to the CO;
$d_{j,CO}$ = the distance between CP and the CO;
e = the cost per meter to connect the CP to the CO.

Using this in the objective function leads to:

$$\min \sum_{j \in J} b_j^* x_j + \sum_{i \in I} \sum_{j \in J} y_{ij} c_{ij},$$

where $b_j^* := b_j + f_{j,CO}$. This keeps the objective function in the same form as Equation 3.1, which means this is again an UFLP and the JMS heuristic can be used.

However, as in the case of FttCab, a star topology will not meet the reliability conditions, which means a ring topology is preferred. Now the connection costs depend on the total set of opened CP locations and the way they are clustered into rings. The first observation also holds for tree topologies. By solving the activation problem of Section 2.2.1 isolated, this is neglected there also. We leave this for further research. However, a simple approach could be to recalculate the connection costs every time a new CP opened, where the connection costs of the closed possible CP locations is based on the shortest distance to an opened CP or the CO.

3.3.2.4.2 Linear or step-wise opening cost function

In the base model the costs of opening a CP where independent of the number of houses that were connected to the CP. There fixed costs were assumed. However, in reality

3.3 FttH planning

the cost will depend on the number of ports or the number of modular elements of the equipment. Normally the opening cost function of location j will be of the form $O_j = b_j + f(k_j)$, using a constant, depending on j and a function depending on the number of houses connected to j. This function can be step wise when there is a modularity in the equipment. In Figure 3.13 an example is shown.

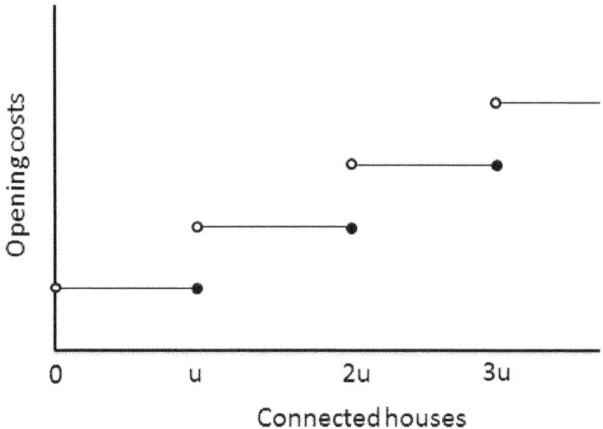

Figure 3.13: Step-wise opening costs.

In that case we deal with a Soft-Capacitated Facility Location Problem (SCFLP) with the opening cost function:

$$O_j = f_j \lceil k_j/u_j \rceil,$$

where u_j is the capacity of one modular system.

The linear variant can easily be calculated by the JMS heuristic, as Mahdian [102] already said: 'Thus, the linear-cost FLP can be solved using any algorithm for UFLP, and the overall approximation ratio will be the same.' This can be seen easily. In the linear case the opening costs are:

$$O_j = b_j + \sum_{i \in I} a_j h_j y_{ij},$$

where b_j is the fixed opening costs, a_j the costs to connect one house, h_i the number of houses in pixel i and y_{ij} the decision variable whether to connect pixel i to CP j. The objective function of the problem then becomes:

$$\sum_{j \in J} O_j x_j + \sum_{j \in J} \sum_{i \in I} c_{ij} y_{ij},$$

where, using Equation (3.3)

$$\sum_{j \in J} O_j x_j = \sum_{j \in J} (b_j + a_j \sum_{i \in I} h_i y_{ij}) x_j = \sum_{j \in J} (b_j x_j + a_j \sum_{i \in I} h_i y_{ij}).$$

This leads to the objective function:

$$\sum_{j \in J} b_j x_j + \sum_{i \in I} \sum_{j \in J} a_j h_i y_{ij} + \sum_{i \in I} \sum_{j \in J} c_{ij} y_{ij},$$

which equals

$$\sum_{j \in J} b_j x_j + \sum_{i \in I} \sum_{j \in J} (a_j h_i + c_{ij}) y_{ij}.$$

This is the form that can be solved by the JMS heuristic.

The SCFLP can be approximated by a LFLP. Again Mahdian [102] showed that the opening cost in the case of a SCFLP can be approximated, for $j \in J$ by:

$$O_j^* = \begin{cases} f_j - f_j/u_j + k_j f_j/u_j & \text{if } k_j \geq 1, \\ 0 & \text{otherwise.} \end{cases}$$

This equals a linear cost function, with fixed costs $f_j - f_j/u_j$ and marginal costs f_j/u_j. Now the SCFLP can be solved by the JMS heuristic, where the total costs are maximal twice the cost of the optimal solution. For the proof of this, see [102].

3.3.2.4.3 Choice of epsilon (ϵ)

The choice of ϵ, which rises the budget in every iteration, is important for the performance of the heuristic. Too small will lead to long calculation times, while there are a lot of iterations needed, in which in most of the cases nothing happens. Too big can give iterations where more options are available and, maybe, the wrong choice is made. The choice of epsilon is there for a balance between accurately and slow, and quick and not that accurately. In the presented heuristic epsilon was fixed for all the iterations. Instead the minimum value of epsilon can be calculated in each iteration and use that. This is the value that triggers the first event, opening of a new CP, at $\epsilon_1^{[t]}$ or assigning a pixel to an already opened CP, at $\epsilon_2^{[t]}$. Now,

$$\epsilon^{[t]} = min(\epsilon_1^{[t]}, \epsilon_2^{[t]}),$$

where each of those two values follow from:

$$\epsilon_1^{[t]} = \min_{J \in F^{[t-1]}} \sum_{i \in N^{[t-1]}} B^{[t-1]} - O_j - c_{ij},$$

and

$$\epsilon_2^{[t]} = \min_{i \in N^{[t-1]}, j \in F^{[t-1]}} B^{[t-1]} - c_{ij},$$

where $N^{[t-1]}$ is the set of not connected pixels after iteration $t-1$, $F^{[t-1]}$ the set of not connected CPs after iteration $t-1$ and $B^{[t-1]}$ the budget after iteration $t-1$.

3.3.3 Connecting the houses

In this section the problem of connection the houses with FttH is studied in more detail and a novel way in modelling this is proposed. Then an algorithm is proposed to solve the problem and then is described how smart co-laying possibilities can be modelled along in this way.

3.3.3.1 Introduction

In this section the problem is considered how to design the fibre paths from the PoP to the individual houses in case of rolling out FttH. As stated before, the problem is considered to be trivial, resulting in comments of constructors like 'The city needs to be opened up completely anyway, right?' However, when observing the problem in more detail it is actually much more delicate and many choices may be made which have a large impact on the final solution and the total costs. Consider Figure 3.14, where we see an example of a street pattern. The single black lines show where fibre optic has to be laid no matter how.

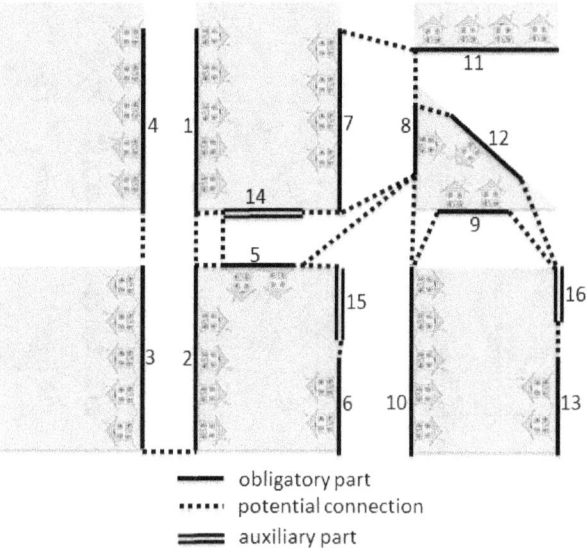

Figure 3.14: Example street pattern.

The double black line segments are auxiliary lines which may be installed if it helps the final solution. For every connection of two line segments there are several alternatives for the manner in which these may be constructed: drilled, dug or laid along in another infrastructural project. Each of these alternatives mean a different price tag for this connection, both for the direct and indirect costs. The possible connections are shown as dotted lines.

Now the line segments need to be pulled together into a coherent network which complies with the following requirements:

1. Each mandatory line segment has to be connected with an access point.

2. The total costs are minimal.

Recall that the costs in the last point may be a combination of direct (equipment, supplies, labour, etcetera) and indirect monetary costs (monetized social costs).

Where all others consider customers as individual nodes, this representation considers connecting line segments which greatly reduces the size of the problem

In order to get a better feeling of the original problem one could view it in a graph representation, see for example Figure 3.15, as introduced earlier. The line segments of the original problem are the nodes in the graph and the possible connections of those segments are the edges of the graph. There are costs assigned to the nodes and edges of this graph. The purpose then is to find a spanning tree within this graph encompassing minimal costs. This is a Minimum Spanning Tree problem as presented in [124]. However, all mandatory nodes (representing the mandatory segments) need to be connected here, whereas the auxiliary nodes are optionally. This is a variant of the minimum Steiner Tree problem where the auxiliary nodes are Steiner nodes.

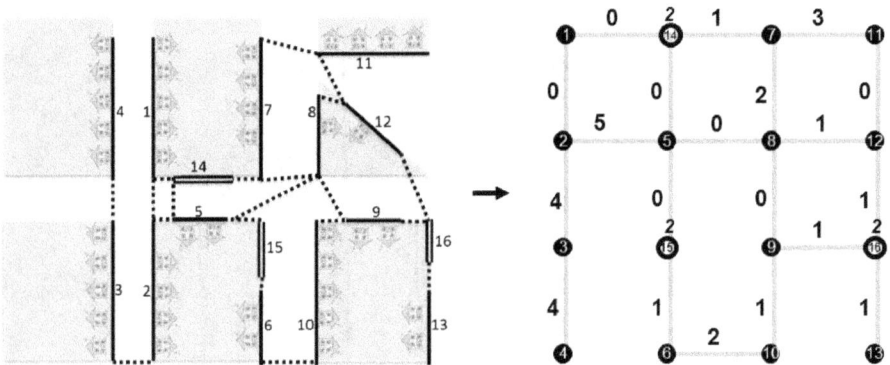

Figure 3.15: Translation to graph representation.

3.3.3.2 Base model

In this section a line segment in the left part or a node in the right part of Figure 3.15 will be referred to as links. Connection will refer to dotted lines in the left part of Figure 3.15 and edges in the right part. In later sections only references to the right part of Figure 3.15 will be made, using the regular terms of edges and vertices.

The first model constructed is the base model. This model searches for a solution of how to connect the links in such a way that the costs are minimized. Here a model is proposed that enables one to connect the access point to all the other targeted points.

The parameters in the models are:

3.3 FttH planning

V : set of n links that need to be connected with the access point;
V^* : set of m links that can be used for connecting with other links;
c_{ij} : the costs to connect the link i to the link j;
d_i : the costs of using link i.

The cost parameter c_{ij} is a combination of the direct costs c_{ij}^{direct} and indirect costs $c_{ij}^{indirect}$. The direct costs consists of the costs to make the connection between link i and link j, like labour costs, cables, equipment etc. The indirect costs are the monetized social costs like lost turn-over, detour costs, damaged greenery etcetera. Now c_{ij} is the weighted combination:

$$c_{ij} = \alpha c_{ij}^{direct} + (1-\alpha)c_{ij}^{indirect}, \qquad 0 \leq \alpha \leq 1.$$

Now solving the model with $\alpha = 1$ gives the cheapest solution for the network owner, the solution with $\alpha = 0$ minimizes social costs. Interaction between network owner or constructor and the municipality should give the optimal vale of α.

When more than one option is available for a connection (such as digging and drilling), each option will have different direct and indirect costs, which can be expressed in the cost function. Option k for the connection of link i and link j has the costs:

$$c_{ijk}(\alpha) = \alpha c_{ijk}^{direct} + (1-\alpha)c_{ijk}^{indirect}.$$

The value of α is known when the model is calculated and then

$$c_{ij} = \min_k c_{ijk}(\alpha).$$

This incorporates the last two options of Section 3.3.1.
The variables that are used for decision-making are, for $i, j = 1, \ldots, n+m$:

$$x_{ij} = \begin{cases} 1 & \text{if link } i \text{ connected to link } j, \\ 0 & \text{otherwise.} \end{cases}$$

$$y_i = \begin{cases} 1 & \text{if link } i \text{ connected to an access point,} \\ 0 & \text{otherwise.} \end{cases}$$

Two auxiliary variables are used to prevent sub tours and for linearisation: u_i and y_{ij}. The model proposed is then:

$$\min \sum_{i \in V \cup V^*} \sum_{j \in V \cup V^*} c_{ij} x_{ij} + \sum_{j \in V^*} y_j, \tag{3.5}$$

under the following conditions:

$$\sum_{j \in V \cup V^*} x_{ij} = 1 \quad (i \in V), \tag{3.6}$$

$$\sum_{j \in V \cup V^*} x_{ij} \leq 1 \quad (i \in V^*), \tag{3.7}$$

$$y_i = \sum_{j \in V \cup V^*} y_j x_{ij} \quad (i \in V^*), \tag{3.8}$$

$$y_i = 1 \quad (i \in V), \tag{3.9}$$

$$u_i \geq u_j - M(1 - x_{ij}) \quad (i,j \in V \cup V^*). \tag{3.10}$$

Expression (3.8) is not linear and can be expressed by introducing y_{ij} as an auxiliary variable:

$$y_{ij} \leq y_j \quad (i,j \in V \cup V^*) \tag{3.11}$$

$$y_{ij} \leq x_{ij} \quad (i,j \in V \cup V^*) \tag{3.12}$$

$$y_i = \sum_{j \in V \cup V^*} y_{ij} \quad (i \in V \cup V^*) \tag{3.13}$$

The conditions can be interpreted as follows:
- (3.5) : Minimize the costs.
- (3.6) : Assign each obligatory link exactly once.
- (3.7) : All the non-obligatory links can be assigned not more than once.
- (3.8) : The link i is connected to the access point if there the link to which it is connected is connected to the access point.
- (3.9) : Each obligatory link must be connected to the access point.
- (3.10) : Prevent the creation of sub tours.

3.3.3.3 Minimum Steiner Tree Problem

The general description of a minimum Steiner Tree Problem is as follows. Given a graph $G = (V, E)$, in which V are the nodes and E the edges of the graph. Make S a subset of V and $c(e)$ a cost function on the sides e from E. We are now interested in de minimal spanning of S in G, in which the costs of the used edges is minimal.

A lot of research has gone into the Minimum Steiner Tree problem. Garey et al. [49] proved that it concerns a NP-hard problem. Kou et al. have taken the initiative for an algorithmic approach in [86]. In [106], Mehlhorn proposes an improvement on this algorithm.

The algorithm of Kou et al. [86] is as follows:

3.3 FttH planning

1. Construct the complete distance matrix $G_1 = (V_1, E_1, d_1)$, with $V_1 = S$ and for every (v_i, v_j) in E_1, $d_1(v_i, v_j)$ being equal to the distance in the original matrix G.

2. Find a minimal spanning tree G_2 in G_1.

3. Construct a sub-graph G_3 in G, by replacing each side in G_2 by the accompanying shortest path in G.

4. Find a minimal spanning tree G_4 in G_3.

5. Construct a Steiner tree G_5 by removing those sides in G_4 which result in the absence of Steiner nodes in the graph endings of the tree.

Waxman [152] shows the performance of an improved version of this algorithm, and of heuristics presented by V. J. Rayward-Smith. He also shows in a simple example that the algorithm does not always determine the optimum, see Figure 3.16.

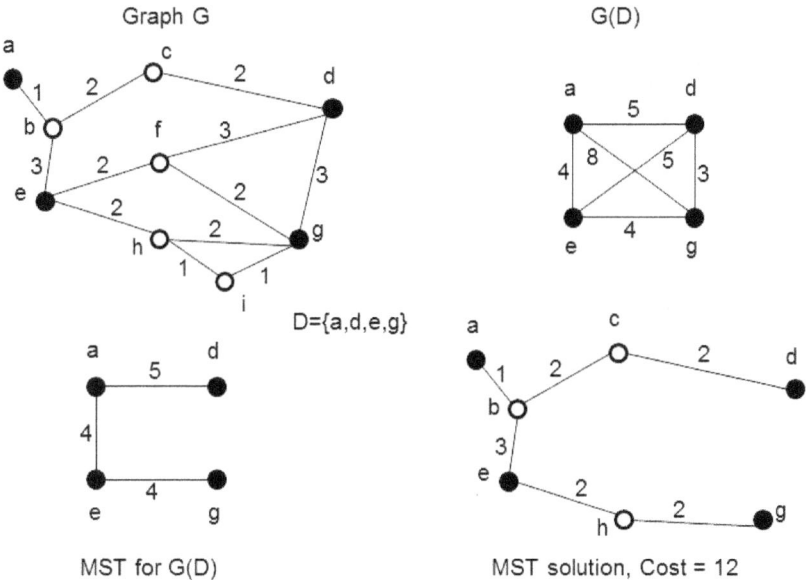

Figure 3.16: Example of Waxman.

The optimum here is the spanning tree, comprised of (a,b), (b,e), (e,f), (f,d) and (f,g), namely $Cost = 11$. The problem in finding this optimum is de dependency of the shortest paths in adding the Steiner points. If Steiner point f is added to the solution, connecting g to the tree costs only 2, instead of 4 (via e) or 5 (via d).

The minimum Steiner Tree also knows variations which may be directly used in network planning. A well-known variant is the 'Prize collecting Steiner tree problem', addressed by a.o. Haouari et al. [60] and an algorithm for the Steiner problem 'with revenue, bottleneck and hop objective functions' by Pinto et al. [96]. Haouari describes a version in which nodes, with the exception of a root node, have been given a price p_j and a penalty γ_j, and a price quota Q. Branches have a cost function. The problem now becomes how to find a subset S and a sub-tree $T(S)$, in which the nodes in S must have a minimum sum Q in prices, and the costs of the branches plus the penalty of the non-selected nodes is minimal. The application of this may be found in the connection of fibre backbone for the connection of a fixed-wireless system. The root node is a particular concentration point, and the other nodes potential locations for antennas of this system. Each of the last locations has a known potential yield (the price) which reflects the attained customers via this location. The penalty reflects the sanction for not installing an antenna in this location. The problem now becomes how to design a network, against minimal connection costs (the costs of the sides) and connecting a subset of locations, which has a higher yield than a pre-set target (Q). In Pinto's article the sides of the graph also have a cost function, and a limited number of nodes also have a yield. The mission now is to find a minimum Steiner tree under the condition of a budget constraint and a hop constraint. As an application example, a fibre optic network installation is given, in which the nodes represent premises to be installed and street junctions. The graph is a representation of an underlying street pattern. The sides of the graph are streets and the cost function a representation of the digging costs between the nodes. The nodes representing premises to be connected have a yield, the sides costs. The availability, or actually the chance of interruptions, is represented by the number of sides between a node and a root node. One might now be interested in a network which maximizes the yield, in which the costs stay below a set budget and every connected node knows a maximum amount of hops (sides) to the root node.

A logical extension to the Steiner Tree problem, that we will need in our problem is the Node Weighted Steiner Tree Problem. Here not only costs are assigned to the edges in the graph but also to the Steiner Points. See for example Klein et al. in [78] and [39] for solution methods. However, this these methods are not designed to integrate the co-laying possibilities.

3.3.3.4 Heuristics

The method presented here is a variation on the method by Kou et al. [86]. However, the whole distance matrix is not determined, like Kou et al. do in step 1, but just the shortest paths which result from removing Steiner nodes. Furthermore, a method is introduced to include the weighted Steiner nodes as edge costs.

The proposed heuristic modifies the target graph in such a way that the Minimum Spanning Tree algorithm can be used.

1. Modification of the graph:

    (a) First, a subset of compulsory vertices (V) and non-compulsory (Steiner) vertices (V^*) is defined. Define the network as a graph $G = (V \cup V^*, E)$, where E is the set of the edges in the graph.

    (b) For each combination of vertices (i, j) in the set V that are connected to each other $(e(i,j) \in E)$ and for which the connection cost c_{ij} is equal to zero: define it as a new vertex

3.3 FttH planning

(c) For all the vertices i in V that are connected with exactly one vertex in V, the following holds: Remove this vertex out of the graph and add the edge to the end solution F, which represents a set of edges.

(d) For all vertices i in V connected with exactly one vertex in V^* and no vertices in V, the following holds: Combine them in a new vertex, where the costs of edges assume the value of the sum of the costs of the edges, increasing them with the vertex costs of the vertex in V^*.

(e) For each vertex i in V^* take each combination of vertices in (j, z) that are connected via i and define the edge (j, z) with costs equal to the costs of the edge c_{ji} increased with the costs of edge c_{iz} and vertex costs of vertex i, d_i.

(f) Remove all the vertices from V^* and remove the edges connecting these vertices from E, denote the resulting graph as $G = (V, E)$.

2. Calculate the Minimum Spanning Tree (MST): Determine the minimum spanning tree in $G = (V, E)$, using Prim's algorithm [124]. This results in the set of the edges F.

3. Translate MST back to original graph: For each edge e in F, determine in the original graph $G = (V \cup V, E)$ the minimum path (using Dijkstra's algorithm [40]) between two end points of the edge e and add this edge to the set F.

4. Improvement step: search in the original graph G for each vertex i in V whether there is an edge in E with x as an endpoint that can substitute an edge in F, with i as an endpoint, so that the solution is improved, and i remains connected.

The last step is needed to assign the costs for the usage of Steiner points to all the outgoing edges. If this vertex is included in the solution, the costs will be accounted one time only.

3.3.3.5 Example

The algorithm described in previous subsection is illustrated with the following example: Assume a network is given as depicted in Figure 3.14. Figure 3.15 shows the graph representation of the same network. The links are represented with vertices and the connections between them with edges. In Figure 3.17(1) we can see that the vertex in the top left corner with 0 costs can be connected to the vertex right below it. The vertex in the top left corner can then be removed from the graph, adapting the edges of the graph in such a way that the existing connections remain unchanged (see Figure 3.17(2)).

If this procedure is repeated for all the edges with 0 costs, the resulting graph becomes as the one shown in Figure 3.18(3). At this point, step 1c of the heuristic described in the previous section is applied. The vertex in the bottom left corner is connected to the graph in only one single way. Consequently, this edge must be added in the final set of edges. Subsequently, we observe that the point above also becomes connected to the graph in a single way. This vertex can also be removed (Figure 3.19). Then, the vertices that are connected to only one (non-compulsory) link are identified. These vertices can be combined in one vertex, resulting in the structure as shown in Figure 3.20(7). It is in

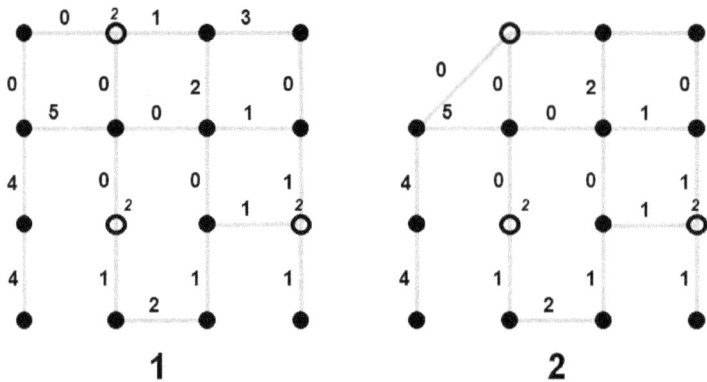

Figure 3.17: Example step 1 and 2.

this graph that we look for Minimum Spanning Tree (step 2). The MST is represented Figure 3.20(8).

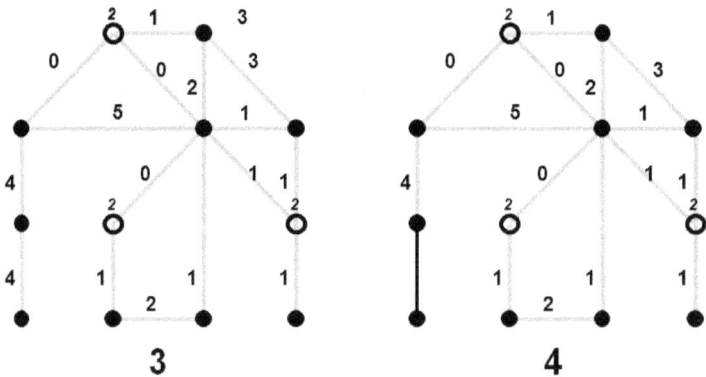

Figure 3.18: Example step 3 and 4.

The next step (3) is to translate the tree back to the original graph. This can be achieved through the execution of the Shortest Path algorithm (e.g. Dijkstra's algorithm [40]). Figure 3.21(9) shows the result of that process. Finally, the improvement step needs to be taken. In Figure 3.21(10) it can be observed that the third vertex from left above can be attached to the graph in another way, offering the savings of 1. The reason this vertex was not directly included in the solution is that in the graph for which the minimum spanning tree has been found, it accounted for the costs of 3. However, including the costs of the non-compulsory links leads to the reduction of extra costs for the new solution of 1. The solution can be translated back to the original network. The final solution is represented in Figure 3.22.

3.3 FttH planning

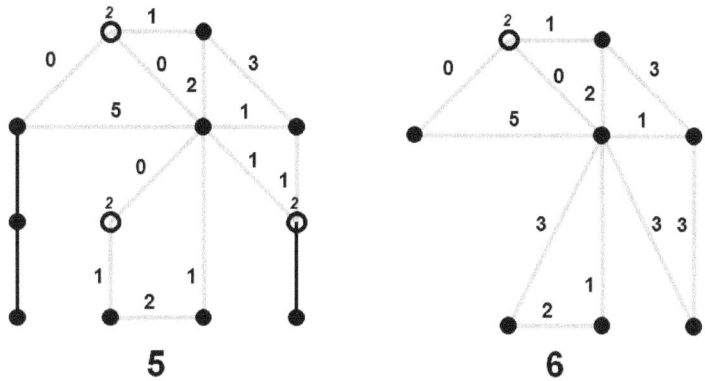

Figure 3.19: Example step 5 and 6.

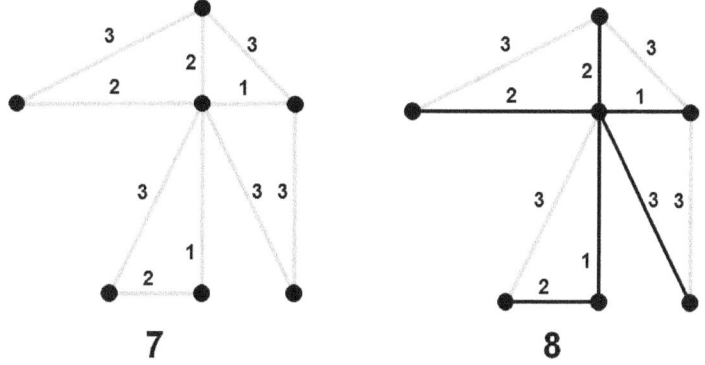

Figure 3.20: Example step 7 and 8.

3.3.3.6 Evaluation

To give an indication of the calculation time and the accurateness of the proposed heuristic we generated some random instances. For various sizes of graphs we generate the costs of the edges, for each size 30 times. Each node (i,j) is connected with nodes $(i+1,j)$, $(i,j+1)$ and $(i+1,j+1)$ if they exist. A fixed percentage of 20% of the nodes is assigned as steiner nodes. This percentage is based on our practical experience. We compare our solution, named $H2$, with the solution of the code created by Fangzhou Chen (www.mathworks.com) based on [53], named $H1$, and with an enumerated exact solution, named Opt. The results are shown in Table 3.4. For the score, the result of $H2$ is set on 100%. $H1$ is a $(2 - \frac{1}{n-1})$ approximation algorithm of $O(n^2 log(n))$. A regular Dutch case of connecting 5000 houses will require around 1000 nodes, 200 of which are auxiliary.

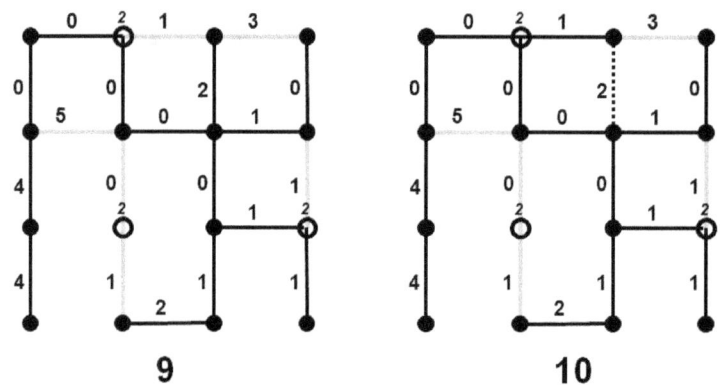

Figure 3.21: Example step 9 and 10.

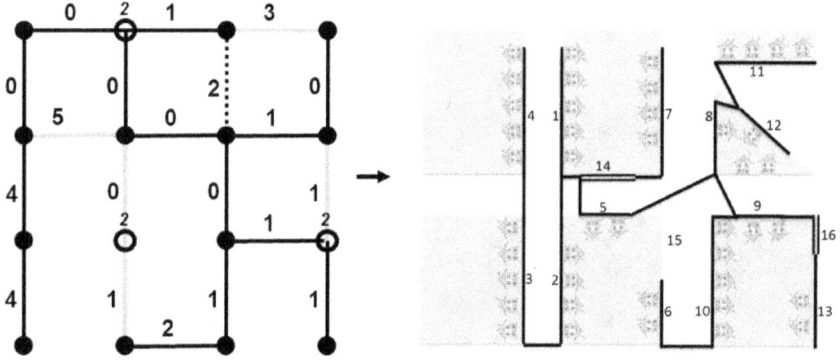

Figure 3.22: Solution in original structure.

3.3.3.7 Model with smart co-laying

In this section we propose an algorithm that also accounts for smart co-laying possibilities, the first option of Section 3.3.1. These smart co-laying possibilities are represented in the graph by assigning to the links the discount on costs and the year in which this co-laying possibility appears. Hence, in addition to the graph $G = (V + V^*, E)$ with accompanying costs $k(x, y)$ where $(e(x, y) \in E)$, there are also the additional parameters $discount(x, y)$ and $year(x, y)$. We now have a Steiner Tree Problem with timing benefits. The algorithm then becomes:

The steps 1a-1f are the same as in the algorithm proposed in the previous section, but an extra item is added in this step:
1 g) The edge costs in the computations that follow are substituted with: $k(x, y) = k(x, y) - discount(x, y)$.
The next steps are now:

2. Denote the resulting graph as $G = (V, E)$. Determine the minimum spanning tree

3.3 FttH planning

Nodes	Score (%)			Time (s)		
	H1	H2	Opt	H1	H2	Opt
25	110	100	100	0.54	0.08	0.02
36	111	100	98	1.41	0.09	0.09
49	118	100	95	3.25	0.10	0.98
64	122	100	93	7.14	0.10	11.01
81	123	100	95	14.92	0.12	126.91
100	120	100	93	29.06	0.13	2842.86
121	122	100	-	67.06	0.16	-
144	126	100	-	121.86	0.18	-
169	127	100	-	213.07	0.21	-
196	124	100	-	322.86	0.26	-
225	125	100	-	587.79	0.31	-
256	127	100	-	965.96	0.41	-
289	125	100	-	1541.27	0.50	-
1000	-	100	-	-	4.89	-

Table 3.4: Performance of the methods.

in $G = (V, E)$. The Minimum Spanning Tree is then determined in the following way:

(a) Given the set F, determine the edge $e(i, j)$ that can be added to the set and that the graph resulting from (V, F) remains connected.

(b) If $e(i, j)$ represents a smart co-laying possibility: Check whether the edge to which $e(i, j)$ is connected is labelled. If the label indicates a year later than $year(i, j)$ select the $e(i, j)$ that has the second-lowest costs, and use it as an input. The procedure splits in two parallel directions:

  i. Add $e(i, j)$ to set F. Label de edge following the $e(i, j)$ with the year $year(i, j)$. Denote it with $F(t)(t = t + 1)$ and continue with the computation of MST for F.

  ii. Remove the discount for $e(i, j) : c_{ij} = c_{ij} - discount(i, j)$. Compute the MST for F.

(c) This results in a set of MSTs F, $F(1)$, $F(2)$, and so forth. Determine the solution with the lowest costs. Denote this with F.

3. For each edge e in F, determine in the original graph $G = (V \cup V, E)$ the shortest path between two end points of the edge e. Add it to H.

4. Improvement step: search in the original graph G for each vertex i in V whether there is an edge in E with i as an endpoint that can substitute an edge in F, with i as an endpoint, so that the solution is improved and i remains connected.

3.3.3.8 Example

The above algorithm is explained in the following example: The start scenario is as shown in Figure 3.23(1). Assume that the connection between the vertices 7-11 has the possibility to be smartly laid down in the year 2015 and that this connection saves the cost of 3. Figure 3.23(1) to Figure 3.23(4) illustrates this process of searching for MST. In step (4), the link with the lowest cost is the link with smart co-laying possibilities. According to the algorithm proposed the problem splits up. There are two MST problems, indicated in Figure 3.24(5) and Figure 3.24(6). Solving these two problems in parallel results in two MSTs shown in Figure 3.24(7) and Figure 3.24(8), respectively. As the solution given in Figure 3.24(8) results in lower costs, it will be this infrastructure that our tool proposes as the most efficient.

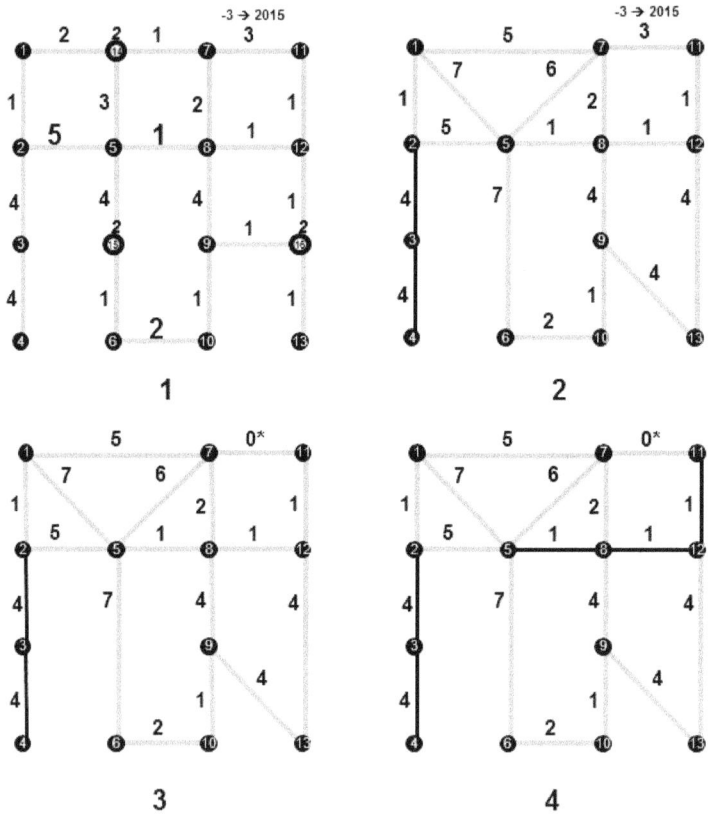

Figure 3.23: Example with smart co-laying step 1-4.

3.4 Economic impact

The fact that we believe that Hybrid FttH is technically feasible and that it might be a bright idea in certain cases, does not mean that it is also feasible from a pure techno-economical point of view. On one hand Hybrid FttH saves cost for installation and for digging into the ground and it may also be quicker to install, but on the other hand it

3.4 Economic impact

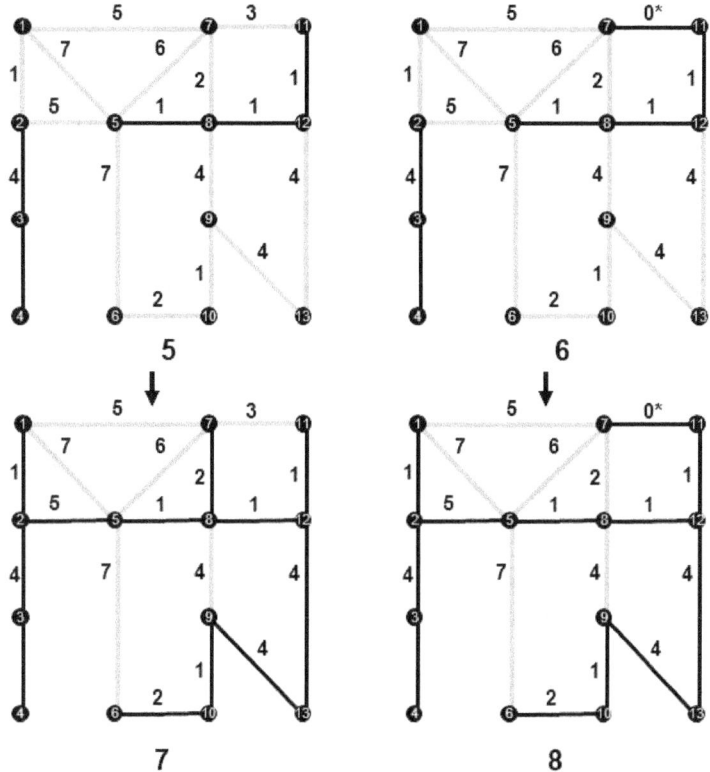

Figure 3.24: Example with smart co-laying step 5-8.

may be more costly in equipment and operational costs. Next to costs the impact of turnover has to be considered. Is it possible to maintain market share in one scenario better than in an other? To this end, a total economical comparison under various scenarios should be made. In this chapter the model that was built for this is presented and demonstrated on two cases.

3.4.1 Top down description of the techno-economic model

In this section an overview of the model that we developed in the CELTIC/4GBB framework [2] is presented. Because of the complexity of some of the underlying mechanisms, the overall model was divided into several sub models. In the next paragraphs we provide a high level overview of the model and describe the sub models and their relations to each other. Our goal is to describe the operations as functional specifications with enough depth that will allow a reader to follow the main logic behind the operations. It is for the moment not our intention to describe each operation in full mathematical detail. For more details see Ahmed et al. [2].

3.4.1.1 Scope

We choose to take an incumbent xDSL network operator as our object of research, and we assume this operator operates a 'VDSL from the CO' access network architecture, the Full Copper variant of Figure 1.1 where all COs are equipped with VDSL equipment. The operator wants to upgrade his clients to a 4GBB-connection. The outcome of the model we propose is the monetary value of applying (A) a certain Hybrid FttH rollout scenario compared to (B) a certain Full FttH rollout scenario. This value is a delta, the difference between (A) and (B). Costs and revenues that do not differ per scenario are not included. We choose for this 'delta' approach based for several reasons:

- In order to calculate the overall profitability of a single roll-out, all cost need to be taken into account, while when looking at deltas, only the costs that differ between the two scenarios need to be examined. This greatly reduces complexity.

- Currently Full FttH is rolled out in a majority of countries in the OECD-world[5], indicating that currently Full FttH is deemed financially viable, at least by the operators involved in the roll out. If we can show that Hybrid FttH is less costly and more profitable than, or at least close to the outcome of Full FttH, we can place the outcome in the total picture. This gives less need to establish a full Profit & Loss statement for a Hybrid FttH roll out.

The second, important, output of the model is the expected market share of the incumbent xDSL operator.

3.4.1.2 Abstract description of the model

A top-down description of the model is given by Figure 3.25. In this paragraph the model is explained from back to front, starting at the output, working our way back through the model and ending up with the input variables necessary to reach this output.
The output of the model is a single monetary value, Delta Net Present Value (DNPV), which expresses the difference in added economic value (EBIT[6]) between two chosen strategies, depicted in scenarios e.g.:

1. Installing Full FttH connections: the Full FttH Strategy

2. Installing a mix of Full FttH and Hybrid FttH connections: the Mix FttH Strategy

The DNPV is calculated by comparing yearly total network investments and costs to yearly total revenues from subscribers between a 'Full FttH' and a 'Mix FttH strategy', adjusting future amounts to their current-day value. This is done in the Output Calculation Sub-Model.

Working backwards from this point, 5 sub-models are used to calculate network cost, connections, subscribers and revenues:

---

[5]The mission of the Organisation for Economic Co-operation and Development (OECD) is to promote policies that will improve the economic and social well-being of people around the world.

[6]EBIT means Earnings Before Interest and Taxes.

3.4 Economic impact

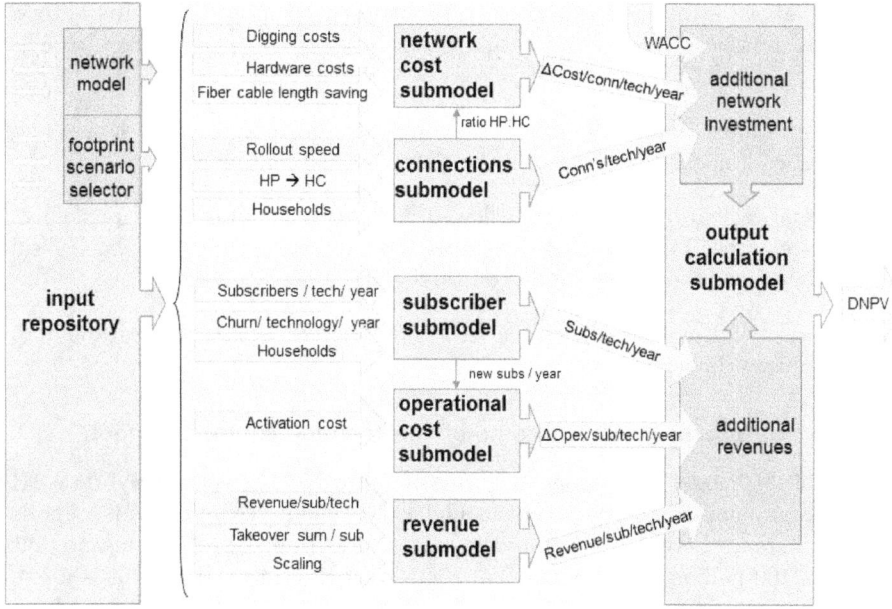

Figure 3.25: Overview techno economic model.

- **Network Cost Sub-model** calculates the total network investments per connection for each year. To do so, it requires numerous network roll-out costs such as digging costs, access network length, labour cost, hardware cost etc. as an input, which it gets from the Input Repository.

- **Connections Sub-model** calculates the number of homes connected to Full FttH and to Hybrid FttH per year. These calculations are based on a set of constraints, such as roll out speed, the convergence from homes passed to homes connected and the total households available, all of which are provided by the Input Repository.

- **Subscriber Sub-model** calculates the number of subscribers per technology per year, using start values of the current market shares of access technologies, and making a yearly redistribution of subscribers over different technologies based on churn figures, as subscribers switch subscriptions between competing infrastructures.

- **Operational Cost Sub-model** calculates the recurring operational cost per technology and activation cost per subscriber per technology.

- **Revenue Sub-model** calculates the marginal revenues for an extra subscriber per year (that is the additional revenues that subscribing one extra customer will yield the operator, given its current cost structure. The revenues per subscriber can be calculated using inputs from company annual reports, recent take-overs as well as calculations based on expert estimations.

Inputs for the above mentioned sub-models are retrieved from the Input Repository. Within the Input Repository, also some sub-sub-models are combined. These models include:

- A Network Model that is used to calculate average distances between homes and access network nodes that is used in the Network Cost Sub-model.

- A Footprint Scenario selector that allows the user to activate several predefined roll-out scenarios. Each scenario prescribes the number of Full FttH and Hybrid FttH connections that will be rolled out in a given year.

- A Scenario Selector that allows the user to select two scenarios with different coverage per technology to be compared.

3.4.1.3 How to interpret the outcome of the techno-economic model

The outcome of this model is the monetary value of applying (A) a certain Hybrid FttH roll-out scenario compared to (B) a certain Full FttH rollout scenario. This value is a delta, the difference between the economic values of (A) and (B). This means that an outcome of '100 million Euro' means that the chosen Hybrid FttH roll-out is worth 100 million Euro more than the Full FttH rollout over a given period of time (at least 10 years). A value of '-100 million Euro' would indicate that a Hybrid FttH roll-out is worth 100 million Euro less than a Full FttH roll-out. This figure does not indicate whether a roll-out as a whole is profitable or loss making, it just indicates the added value of the one roll-out option over the other. Still, calculating the difference between the two scenarios greatly relies on interpretation of the model. Therefore the results of this model first and foremost have to be considered as a first estimation of the value of the Hybrid FttH concept, in order to strengthen technical developments, and only second as a basis for investment decisions. The Hybrid FttH concept is still very much a work-in-progress, meaning that the outcomes of this model will be subject to change based on future (technical or economical) insights. Next to a Delta NPV figure, which is the end-outcome of the entire financial model, the outcomes of the sub-models, such as installation costs or number of homes connected per year (roll-out speed) are also interesting for the further development of the Hybrid FttH discussion and can be used for case by case comparison for streets or (parts of) cities to determine which technique is more economical.

3.4.2 Sub-models in detail

In the previous section the 5 sub-models were presented. In this section these sub-models are presented in more detail.

3.4.2.1 Network cost sub-model

The Network Cost Sub-model calculates the total network investments per connection for each year. For the calculations of the cost we assume a certain network roll out. The structure of this roll out is sketched, the specific figures are an example.

3.4 Economic impact

3.4.2.1.1 Assumptions

For the calculations of the cost of each topology migration path we assume a certain topology roll out, with choices regarding technology, structure, dimensioning etcetera. We assume that a G.Fast node has a capacity of X_3 connections, that X_2 G.Fast nodes can be connected to one cabinet and that X_1 cabinets can be connected to one ring to the CO. This makes the total connections on a cabinet X_2 times X_3. This is shown in Figure 3.26. For example, a FttCab fibre ring has a maximum of 2500 connections, divided in 6 cabinets, each with capacity of 384 connections, connected by a fibre ring. When the migration from FttC to Hybrid FttH is performed, each cabinet gets 8 Hybrid FttH nodes, each having 48 connections, connected in a star structure. The fibre ring can be fed by one FttH PoP, the location of this PoP is already determined and taken into the ring. This is shown in Figure 3.26.

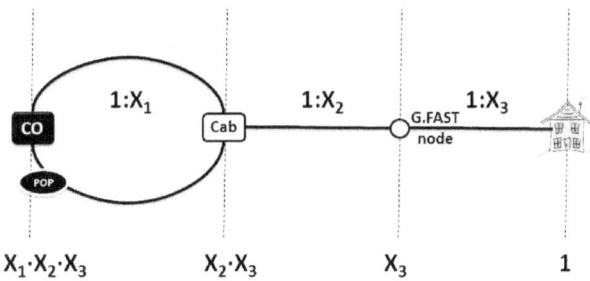

Figure 3.26: Design choices of the network.

3.4.2.1.2 Geometric model

For the cost model we need to calculate distances of both trench and cable. Next to that, we like to know something about the (expected) maximum distance in the roll out. For these calculations we use a geometric model, as also used in e.g. [22]. We distinguish four main variants in this geometric model:

1. Star structure, single sided houses.

2. Star structure, double sided houses.

3. Snake structure, single sided houses.

4. Snake structure, double sided houses.

These four structures are shown in Figure 3.27. We define A as the access point, n as the number of houses to be connected, $s = \sqrt{n}$ and d = width of the premises. For each of these structures the length of the trench and cable (without gardens) can be calculated:

1. Star structure, single sided houses: distance of digging $= d \cdot (s+1)(s-1)$, distance of cable $= 2 \cdot d \cdot s \cdot \lceil \frac{1}{2}s \rceil \cdot \lfloor \frac{1}{2}s \rfloor$.

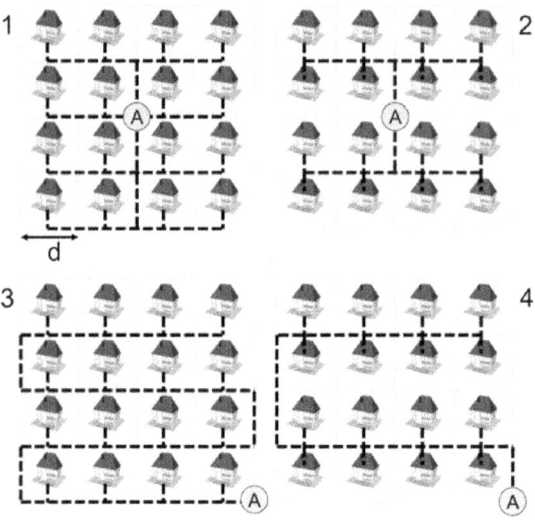

Figure 3.27: Four geometric models.

2. Star structure, double sided houses: distance of digging = $d \cdot (\frac{1}{2}s \cdot (s-1) + s - 2)$, distance of cable = $2 \cdot d \cdot s \cdot \lceil \frac{1}{2}s \rceil \cdot \lfloor \frac{1}{2}s \rfloor$.

3. Snake structure, single sided houses: distance of digging = $d \cdot ((s^2 - \frac{1}{2}) + (s-1))$, distance of cable = $d \cdot (\frac{1}{2}s^3 + \frac{1}{2}s^2(s^2 - 1))$.

4. Snake structure, double sided houses: distance of digging = $d \cdot ((\lceil \frac{1}{2}s \rceil \cdot s - \frac{1}{2}) + (s-1))$, distance of cable = $d \cdot (s^2 + \frac{1}{2}s^3 + K \cdot s \cdot (s+2)) \approx d \cdot \frac{1}{4}(s+2) \cdot s^3$, where $K = \frac{1}{2}(\lceil s/2 \rceil) \cdot (\lceil s/2 \rceil - 1) + \frac{1}{2}(\lfloor s/2 \rfloor) \cdot (\lfloor s/2 \rfloor - 1)$

The maximum copper distance can be calculated by:

1. Star structure, single sided houses: $s \cdot d - 0.5d$.

2. Star structure, double sided houses: $2\lceil \frac{1}{2}s - 1 \rceil \cdot d + 0.5d$.

3. Snake structure, single sided houses: $s^2 \cdot d + s \cdot d - 1.5d$.

4. Snake structure, double sided houses: $\lceil \frac{1}{2}s \rceil \cdot s \cdot d + 2 \cdot \lceil \frac{1}{2}s \rceil \cdot d - 1.5d$.

For a multi layer network, each layer can be treated as a separate geometric model. In the architecture as shown earlier three layers can be distinguished:

1. The CO as the access node and the 8 cabinets to be connected.

2. The cabinet as access point and the 8 Hybrid FttH nodes to be connected.

3. The Hybrid FttH node as the access node and the 48 houses to be connected.

In this analysis we assume for layer 1 a ring structure, for layer 2 a star structure, single side, and for layer 3 a snake structure double sided, as seen in some European countries like The Netherlands.

3.4 Economic impact

Description	costs	unit
Digging, closing trench	15	€/m
Breaking and repairing tiles	10	€/m
Fibre (Direct buried cable)	0.3	€/m
Drilling (garden)	25	€/m
Duct	2	€/m
Blowing fibre or cable	500	€/duct
Hybrid FttH node E & I	2500	€/node
Cabinet E & I	11000	€/node
Premises E & I	250	€/node
Removing equipment	250	€/node
End user equipment	100	€/connection

Table 3.5: Cost input parameters.

3.4.2.1.3 Cost Parameters
We distinguish the following cost categories:

1. Connection CO to cabinet: Digging, closing trench, breaking and repairing tiles; ducts.

2. Equipment and (de-)installation cabinet.

3. Connection cabinet to Hybrid FttH node: Digging, closing trench, breaking and repairing tiles; ducts.

4. Equipment and (de-)installation Hybrid FttH node.

5. Connection Hybrid FttH node to premises: Digging, closing trench, breaking and repairing tiles; direct buried cable.

6. Equipment and (de-)installation in premises.

The used values are in Table 3.5.[7]

3.4.2.1.4 Validation
For a rough validation we look at the results of Section 3.2. Here we calculated the cost for two cities in the Netherlands, Amsterdam and The Hague, in detail, using the activation algorithm of Section 2.2.1 and Prim's algorithm [124]. We assume that the cabinets already have a fibre connection, so our focus is the part of the network between the cabinet and the home connection. The Amsterdam case has 150,058 branching point, The Hague 89,076. Those branching points are the potential spots to place the new equipment. In both cities we want to connect at least 99% of the customers within 200 meter to a G.Fast node. Each G.Fast node is placed in a manhole. We can place each combination of 16-port and 48-port G.Fast equipment (G.Fast multiplexer) in the manhole. Now we plot the resulting costs per FttCab connection of the various COs and their average connection density in Figure 3.28 and compared it with the results of

---
[7]These values come from the TNO cost database, filled by input of various constructors

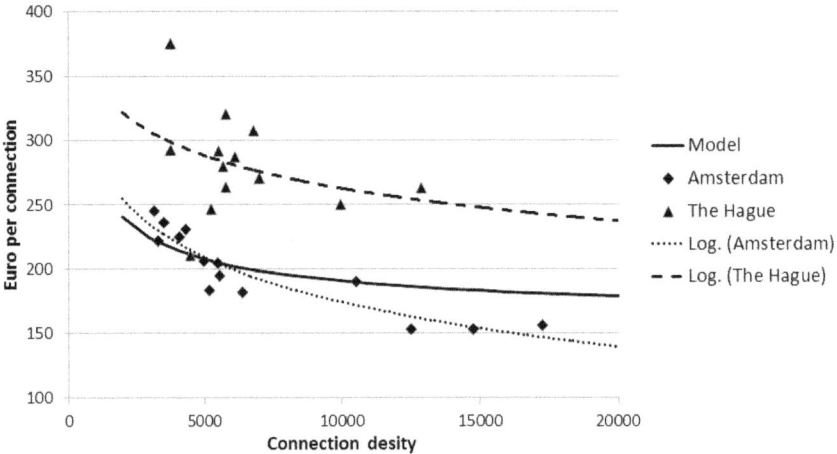

Figure 3.28: Validation of the model.

the simple geometric model. For both cities we plot a logarithmic trend line to indicate the underlying relationship. The differences between Amsterdam and The Hague follow from the size of the cabinets. In The Hague the current cabinet size is bigger; this increases the average length between the cabinet and the new activated points.

3.4.2.2 Connections sub-model

The Connections sub-model calculates the number of homes connected to Full FttH and to Hybrid FttH per year. These calculations are based on a set of constraints, such as roll out speed, the convergence from homes passed to homes connected and the total households available, all of which are provided by the Input Repository.

The Connections sub-model takes several input parameters to calculate the number of homes connected (=homes that can actually subscribe to full/Hybrid FttH) per year, including:

- **Roll-out capacity (R)**: There are several constraints that limit the maximum Full FttH roll-out capacity. In our model, we assume that Full/Hybrid FttH is rolled out at maximum capacity each year, that is controlled via a parameter.

- **Households**: The number of households in the target area. New homes are built, old homes are demolished, people marry, people divorce. This leads to a changing number of households each year. The model logically assumes that no more FttH can be rolled out than the number of households available.

- **Construction time ratio (CTR)**: Another crucial assumption is that we assume that the construction of Hybrid FttH takes less time than the construction of Full FttH, because Hybrid FttH uses the already pre-existing copper access network into the customer premises. From previous Full FttH roll-outs we have learned that most time and effort goes into connecting the last 50 meters of fibre onto

3.4 Economic impact

the customer's premises. The construction time ratio is a number expressing the number of Hybrid FttH connections that can be constructed in the time it takes to construct 1 Full FttH connection (homes passed). If this ratio for instance is 4, that would mean that instead of constructing 1.000 connections of Full FttH, an operator can in the same time construct 4000 connections of Hybrid FttH.

- **Hybrid FttH households ratio**: This ratio expresses the number of households suited for Hybrid FttH, as a percentage of the total number of households in the target area. We expect Hybrid Fibre to be more suited for densely populated areas, where people have no garden, live in stacked buildings up to a certain number of stores. We also assume that operators will connect buildings to the most suited type of access network.

- **Minimal Full FttH ratio (MFR)**: This ratio expresses the minimum number of connected Full FttH connections, related to the connected Hybrid FttH connections.

- **Homes passed vs. homes connected ratio (HCR)**: When homes are connected to Full FttH, several states of connectedness exist:

  - Homes passed: fibre is constructed up to the premises, but not connected inside the home. In order to construct a 'home passed', no interaction with the owner of the premises is required, but in order to construct a 'home connected', a home owner has to be involved to give (written) permission, and allow the construction crew access to the building. In previous roll-outs, a percentage of up to 10% of homes passed have never been transformed into home connected. Some reasons were unresponsiveness (away on a long holiday, locked up in jail, deceased), or objected to getting the connection in general.

  - Homes connected: fibre is installed up to the premises, and guided through one of the walls of the home, and attached to an ONT (Optical Network Termination point) inside the home. The ratio used in our model is the percentage of homes passed that within a year are upgraded to homes connected. Hybrid FttH does not suffer this drawback, because the existing copper line into the customer's premises is used. Therefore, the homes passed versus homes connected ratio for Hybrid FttH is 1.0.

The model calculates two scenarios: a Full FttH roll-out and a mixed roll-out (Full FttH and Hybrid FttH). If x is the number of homes passed with Full FttH, a maximum number of $(R-x) \cdot CTR$ Hybrid FttH connections can be created. This is the remaining roll-out capacity $(R-x)$ times the higher speed to roll-out Hybrid FttH (CTR). The connected number of Full FttH connections must be more than the fraction MFR of the total connected connections. This results in the constraint:

$$HCR \cdot x \geq MFR \cdot ((R-x) \cdot CTR + HCR \cdot x)$$

This results in

Input parameter	Values
Rollout capacity (R)	100,000
Construction time ratio (CTR)	400%
Hybrid FttH household ratio	20%
HP vs. HC ratio Full FttH (HCR)	90%
Minimal Full Fibre ratio (MFR)	50%
Result Full FttH Scenario:	
Homes Passed Full FttH	100,000
Homes Connected Full FttH	90,000
Result Mixed Scenario:	
Homes Passed Full FttH	81,632
Homes Connected Full FttH	73,469
Homes Connected Hybrid FttH	73,469
Homes Connected total	146,938

Table 3.6: Scenario generating example.

$$x \geq \frac{MFR \cdot R \cdot CTR}{HCR + MFR \cdot CTR - MFR \cdot HCR}.$$

In Table 3.6 you find an exemplary calculation showing that the total number of homes connected for each scenario is different. The capacity of the roll-out is 100,000 connections. If we use this capacity for Full FttH, we can pass 100,000 houses and connect 90,000 of them. If we introduce Hybrid FttH, we have to find the number of passed houses with the given formula. This gives 81,632 houses passed with Full FttH, of which 73,469 are connected. The same number of Hybrid FttH connections can be realized with the given roll-out capacity. A total number of 146,938 houses are connected with a 4GBB connection.

Next to generating this number automatically using the parameters, the user has also the possibility to describe two scenarios directly, without using this model.

3.4.2.3 Subscriber sub-model

The Subscriber Sub-model calculates the number of subscriber per technology per year, using start values of the current market shares of access technologies, and making a yearly redistribution of subscribers over different technologies based on churn figures, as subscribers switch subscriptions between competing infrastructures.

The reasoning behind the working of this sub-model is based on the assumption that customers will churn to an access network that offers the highest bit rate (at a similar price).

3.4.2.3.1 Used services

In order to estimate the number of subscribers over the years, the model first needs a start set of current subscribers per competing broadband access technology category. Categories used are:

3.4 Economic impact

- xDSL: All infrastructures using a copper access network of more than 200 metres from the home. This includes all flavours of ISDN, ADSL, VDSL from the CO and VDSL from the cabinet (also referred to as FttCab).

- Full FttH: All infrastructures consisting of 100% fibre up to the customers home

- Hybrid FttH: All infrastructures using a copper access network at a range of 20 to 200 metres from the home. This can also be referred to as FttBuilding, FttCurb, etc.

- Competition: predominantly HFC (Hybrid FttH Coax): all infrastructures using a coaxial access network, but also mobile connections (4G, 5G, WiMax), satellite connections etcetera.

In this chapter we define churn as a customer going form xDSL ISP[8] A to HFC ISP C, not a customer going from one xDSL ISP to another xDSL ISP. Also note that the number of customers per access technology is determined using broadband internet service as the dominant factor, not TV service or telephony service. Customers can, for instance, have a xDSL connection for internet service, and a HFC Coax connection for (digital) TV. We count subscribers based on the fixed line connection on which a household receives broadband internet services.

3.4.2.3.2 Areas

Next to knowing how many customers have subscribed to an internet service using a certain access technology, the model also needs figures on the total availability of each access technology. By availability we mean the total number of households that have a physical connection to an access network, not the number of households that actually subscribe to it. For instance, ADSL in The Netherlands is available almost everywhere - for this model we assume 100%. That means that all households in The Netherlands could subscribe to an xDSL service if wanted.

We divide the country into several areas, each with a different availability of technologies:

- only xDSL infrastructure;

- only xDSL and Competition infrastructure;

- only xDSL and Full FttH infrastructure;

- only xDSL and Hybrid FttH infrastructure;

- xDSL, Full FttH and Competition infrastructure;

- xDSL, Hybrid FttH and Competition infrastructure.

---
[8]Internet Service Provider

		From: xDSL	FttH	Competition
To:	xDSL	60%	5%	8%
	FttH	20%	85%	12%
	Competition	20%	10%	80%

Table 3.7: Churn figures year 1 area 1.

We assume xDSL has a 100% coverage, so no areas exist that contain only Competition and Full or Hybrid FttH infrastructures exist.

The model calculates the total households that fall within each area for each year, based on inputs from technology roll-out estimations that come from the input preparation model.

3.4.2.3.3 Starting values

The model also requires figures for the number of households within the target area for the starting year and household developments over time. We also need internet penetration, which is how many people use internet services. These figures are estimated by government agencies such as the National Statistics Offices. Combination of these two figures allows us to calculate the total addressable market for broadband internet connections. Thus far, households and internet penetration have both grown each year, leading to a double growth (more households, higher percentage of those households) of the addressable market for broadband internet, but in the future. This growth seems to be slowing down as we approach 100% penetration combined with an ageing and possibly shrinking population.

3.4.2.3.4 Churn

In order to complete our estimation of future subscribers per access technology, we need to take into account that customers switch from one type of access technology to another. For each year, the model will calculate the flows of customers between the different technologies, using churn figures per technology. The percentage of customers that leave a technique in a year is called the churn percentage. Current churn figures can be deducted from public sources such as operator's annual reports, but future churn is subject to interpretation. The churn figures used for this model can be based on estimations of several experts, but do have a high degree of uncertainty. Later on we show also a method to construct the churn figures by a model.

We assume churn is affected first and foremost by (lack of) sufficient bit rate. When confronted with a higher bit rate at a similar price, a certain number of customers will switch. In reality, many other factors such as pricing, marketing or economic development determine the actual churn/gain.

For each access technology, a list of churn figures per year is required. In Table 3.7 is an (indicative) input for an area in which xDSL, Full FttH and Competition networks are available. There we see churn figures for Area 1, in year 1. This area has both xDSL, FttH and Competition (Cable, 4G) network infrastructure present, and customers will switch between these three infrastructures. In year 1, out of all customers who

3.4 Economic impact

Service	Subscribers
xDSL	200
FttH	10
Competition	100

Table 3.8: Subscribers area 1 year 1.

		From: xDSL	FttH	Competition	Total EoY
To:	xDSL	120	1	8	129
	FttH	40	9	12	61
	Competition	40	1	80	121

Table 3.9: Churn results area 1 year 1

currently have xDSL, 20% will leave to get an FttH subscription, and another 20% will leave for a Competition subscription[9]. The churn for xDSL is (20%+20%=) 40%. Customers from other networks also churn towards xDSL. From all FttH customers, 5% churn towards xDSL. From all Competition customers, 8% churns towards xDSL. We need churn figures for these areas for each year, since churn figures can change over time due to developments in the (relative) bit rate of the different technologies.

3.4.2.3.5 Calculation

This paragraph explains the operations within the Subscriber Sub model that transform the inputs (churn figures, subscriber start values, connection availability, household developments, etcetera) into the output of subscribers per technology per year.

We will go through a step-by-step (simplified and hypothetical) calculation for a single area in order to demonstrate the working of our model.

Assuming we start with the subscribers shown in Table 3.8. Households within this area who have no internet connection are not shown here. When multiplying the churn figures in Table 3.7 with these subscribers figures, we get the results as shown in Table 3.9. Here EoY stands for the End of Year figures.

This means that in the end of year 1, xDSL has lost 80 customers, while Ftth has gained 51, and Competition has gained 21.

Assuming households (combined with internet penetration) grow at a rate of 5%, and new subscriptions are divided equally among these households (which is a simplification that is not being used in our actual model), we start year 2 with 134 xDSL, 64 FttH and 127 Competition subscribers, following from Table 3.10. Here BoY stands for the Begin of Year figures.

Assuming that for year two the same churn figures as for year one apply, and than the same growth in households, the same churn, and so on, we end up with the overview of subscribers at the first four years, as shown in Table 3.11.

---

[9]This does not mean xDSL will have (60%+5%+8%)=83% of its original subscribers left after one year, because the FttH and Competition churn are not relative to the number of xDSL subscribers, but to the number of FttH or Competition subscribers.

Service	EoY 1	new	BoY 2
xDSL	129	5%	135
FttH	61	5%	64
Competition	121	5%	127

Table 3.10: Subscribers year 2.

Service	BoY 1	BoY 2	BoY 3	BoY 4
xDSL	200	134	99	80
FttH	10	64	101	129
Competition	100	127	142	150

Table 3.11: Subscribers BoY 1 to 4.

We can see clearly that almost half of the original subscribers of xDSL have churned away. FttH has grown to be the technology with most subscribers, while Competition has also been showing healthy growth.

Figure 3.29 and Figure 3.30 show the same decline of xDSL, the fast growth of FttH, and the modest growth of Competition. The first Figure for a growth of 5% as in the example, the second for a growth of 0%. One other property of this graph is that all lines seem to slope towards an equilibrium. xDSL will never reach 0, FttH will never reach 100% market share. In the case the churn matrix is time independent, this equilibrium can be calculated easily using the theory of Markov chains, see for example [144] for theory about Markov Chains, and [130] for application in marketing. Let the vector x_t be the vector of subscribers at time t. let P be the churn matrix of Table 3.7. Then the number of subscribers of all n services at time $t+1$ follows from $x_{t+1} = Px_t$. The stationary distribution π of the subscribers follow from $\pi = P\pi$. To solve this, one have to solve $n+1$ equations:

$$\pi_j = \sum_{i=1}^{n} \pi_i P_{ij} \quad j = 1, \ldots, n,$$

$$\sum_{i=1}^{n} \pi_i = 1.$$

If P is irreducible[10] and positive recurrent[11] then it has an unique stationary distribution, see for example [144]. The equilibrium function (E) of service i in year t is then calculated by:

$$E_{it} = T\pi_i(1+r)^t,$$

where T is the total sum of subscribers at $t = 1$, and r the growth rate of the households.

The calculations demonstrated in this chapter are a simplification of the actual calculations within the Subscribers Model. Some differences are:

---
[10] A Markov chain is said to be irreducible if a state i is reachable from any other state j.
[11] A Markov chain is said to be positive recurrent if all states have a finite expected return time

3.4 Economic impact 107

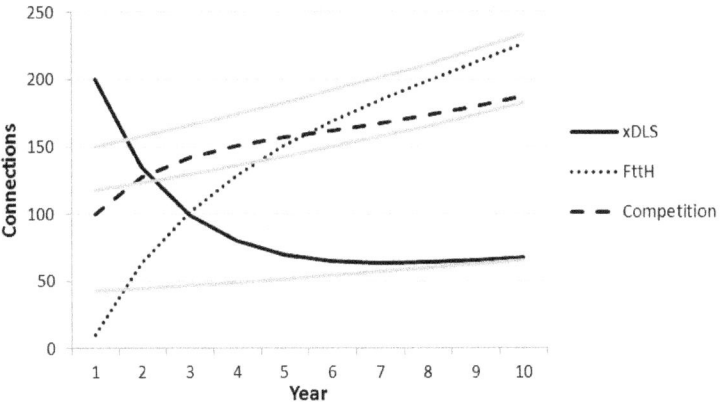

Figure 3.29: Effect of churn on market share, $r = 0.05$.

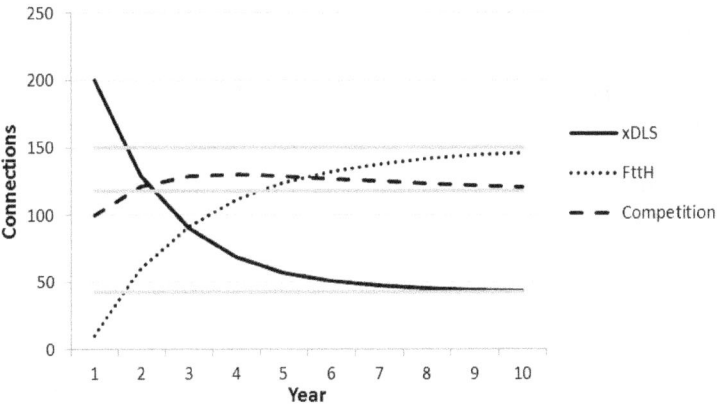

Figure 3.30: Effect of churn on market share, $r = 0.00$.

- The model we have constructed combines these churn-calculations for all areas, adjusting the area-size each year to accommodate for the roll-out of both Full and Hybrid FttH.

- Not only churn, but also uptake curves for new technologies (Full and Hybrid FttH) is taken into account

3.4.2.4 Generate churn by model

An alternative approach is to generate the churn figures by a model. Assume we have a distribution function that describes the need for broadband access over the population. Next to that, we assume that this distribution function follows a normal distribution with density function $f(x)$, parameters μ and σ. Define $P(b)$ as the fraction of consumers that has a need for broadband access with download speed b Mbit/s, then:

$$P(b) = \begin{cases} \int_{b-0.5}^{b+0.5} f(x)dx & b > 1, \\ \int_{-\infty}^{0.5} f(x)dx & b = 0. \end{cases}$$

To what proposed service is a (potential) customer with need b attracted? We model that using gravitation theory from physics. According to the law of universal gravitation, the attractive force (F) between two bodies is proportional to the product of their masses (m_1 and m_2), and inversely proportional to the square of the distance, r (inverse-square law), between them:

$$F = G\frac{m_1 m_2}{r^2}.$$

The constant of proportionality, G, is the gravitational constant. We here define

$$F(s,b) = \frac{1}{(s-b)^2}$$

for the attractive force of a service with speed s to a customer with broadband need b. Next we calculate the total force of that service by taking

$$F_{total}(s) = \sum_{b=0}^{\infty} F(s,b) * P(b).$$

We now assume that the long term market share of this service is equal to the fraction of that forces:

$$M(s) = \frac{F(s,b)}{\sum_{s' \in S} F(s',b)}.$$

However, the need for broadband access is not constant, we have seen it growing the last decades and we assume it will be growing for some years. TNO [146] estimates that the bandwidth demand between now and 2020 on fixed connections will grow exponentially by approximately 30% to 40% per year. Figure 3.31 comes from this report and depicts the growth in average need for bandwidth and the spread in bandwidth demand. This spread follows the distribution function, so we get a μ_t and σ_t and we can calculate $M_t(s)$ for all years and services. This long term market share will not be realized in one year, but we assume in this year the market will start acting towards this long term situation. Mathematically we now have the opposite situation as earlier, where we were looking for the stationary distribution from a Markov chain. We now start with a stationary distribution and want to use the corresponding Markov matrix as the churn matrix for the year concerned, so M_t is de left eigenvector of P with eigenvalue 1 and P is a right stochastic matrix with non-negative values, such that $M = MP$.

We show this in the following example. We are again in the area that has both xDSL, FttH and Competition. Let us assume that the xDSL has a bit rate of 10Mbit/s, Competition 30Mbit/s and FttH 80 Mbit/s. Now we can calculate for all services the

3.4 Economic impact

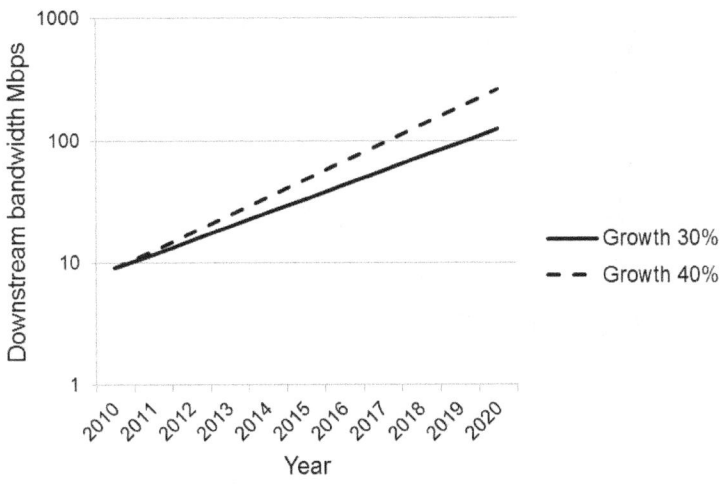

Figure 3.31: Expected broadband demand (source [146]).

long term market share $M(s)$ for all possible values of μ and σ. This is shown in Figure3.32. Next, we can calculate the long term market share per year $M_t(s)$ for the expectation following from Figure 3.31. These are shown in Figure 3.33. The demand in this Figure is the average demand from Figure 3.31.

Next we have to solve the set of equations $M_t = M_t P_t$ for each year. Unfortunately this set of equations do not have a single solution. There are some degrees of freedom. We choose to fill these degrees freedom all with the same value x. Then there is a range $(0, a)$ this value can take, such that no other value of the matrix gets a negative value. If you choose 0 you will get the identity matrix (or unit matrix), the matrix with ones on the main diagonal and zeros elsewhere. With x close to one, the convergence to the stationary distribution will be very slow, with x close to a the convergence will be very fast. In this example we choose $x = \frac{1}{2}a$. This results in the realized market shares as shown in Figure 3.34.

3.4.2.5 Operational cost sub-model

The Operational Cost sub-model calculates the recurring operational cost per technology and activation cost per subscriber per technology.

In order to calculate the marginal profit for a single business case, one way to proceed is to have all cost of the company available. For this exercise however, that is an impossible task, because we would need to get our hands on classified information from existing companies. Luckily, because we do not want to build a complete business case, but only want to compare scenarios, we only need to compare those cost and revenues that differ between the Full FttH scenario and the Mix FttH scenario. Public sources and annual reports give us some insights into an average telecom companies cost structure. We define two types of operational cost: Recurring operational cost and one-off activation cost:

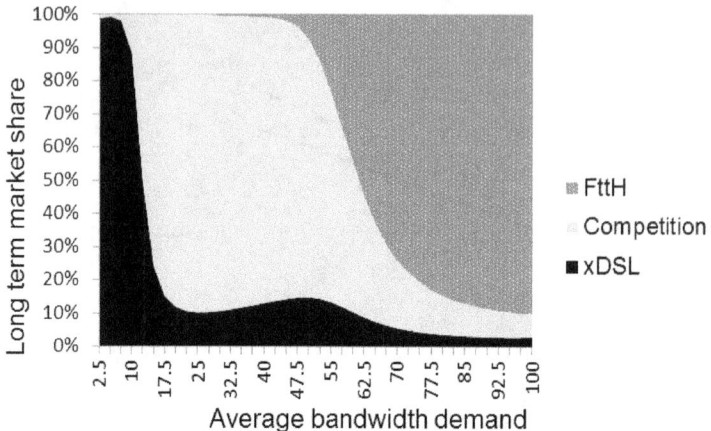

Figure 3.32: Long term market share for different average broadband demand.

Revenues:	100 x 250	= 25,000
Network cost:	1 x 10,000	= 10,000
Operational cost:	100 x 100	= 10,000
Profit:	25,000-20,000	= 5,000

Table 3.12: Business case part 1.

- Recurring operational cost: The operational costs put in the input sheet are multiplied by the number of subscribers per technology, giving a total operational cost per technology. These total numbers are transferred to the business case.

- One-off activation cost: Activation cost put in the input sheet are multiplied by the number of activations per year. This number is calculated by taking the number of churning customers (calculated by the Markov model) and adding the rehousing customers (put in the input sheet as percentage rehousing). As mentioned earlier, for model simplicity sake, we assume rehousing customers will keep their current subscription, but similar activation cost are made as a new customer.

3.4.2.6 Revenue sub-model

This model calculates the value of an additional subscriber in a given year. In economic terms this is referred to as marginal revenue. It is defined as the additional revenue that the production of one extra unit of product will bring.

To understand the concept of marginal profits for a telecom company, imagine an operator with 100 customers. The company spends a total of 10,000 Euro on network investments each year. The company also spends 100 Euro per customer on operational costs (billing, service desk, etcetera). Each customer pays 250 Euro per year.

The overall (simplified) business case of the company will look like shown in Table 3.12. The profit per customer is 5,000 Euro divided by 100 customers equals 50 Euro

3.4 Economic impact

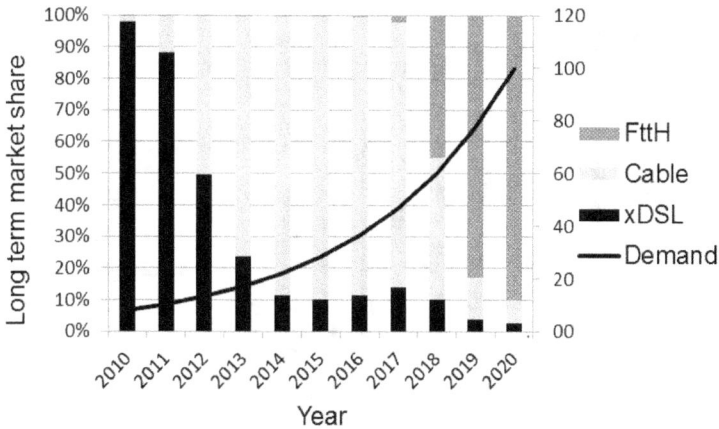

Figure 3.33: Long term market share per year.

Revenues:	101 x 250	= 25,250
Network cost:	1 x 10,000	= 10,000
Operational cost:	101 x 100	= 10,100
Profit:	25,250-20,100	= 5,150

Table 3.13: Business case part 2.

per customer. Imagine the company attracts one new customer. That would make the business case look shown in Table 3.13. The profit per customer is now 5.150 divided by 101 customers equals 51.50 Euro per customer. However, when determining the value of the single new customer, we notice that by attracting him/her, the company has gained 150 Euro in profit. Therefore, the marginal revenue of a customer is 150 Euro. This is because the network costs are fixed. The network cost are 10,000 Euro a year, regardless of being used by only one customer or by a 1,000.

Telecom companies usually have high fixed cost, therefore it is very lucrative to attract additional customers, to be able to divide the large shared network cost over as many subscribers as possible. On the other hand, this makes telecom companies also at risk of a downward spiral when loosing too many customers. The fixed network costs stay the same, so the remaining customers need to pay more on a per-customer basis to keep the network up and running, leading to a higher churn, leading to higher costs, leading to higher churn, etcetera.

The Revenue sub-model calculates this marginal revenue per customer. As input variable we need the revenues per customer, also known as ARPU (Average Return Per User). These figures can be calculated using information from company annual reports. The ARPU for a FTTH subscription is higher than the ARPU for an xDSL subscription.

Another input we need is information on the scaling of the company. Returning to the example of the previous paragraph, this process of adding customers cannot go on indefinitely. At a certain moment in time, the network will be overloaded, and needs to be upgraded. The cost for the network are not totally fixed, but increase at certain

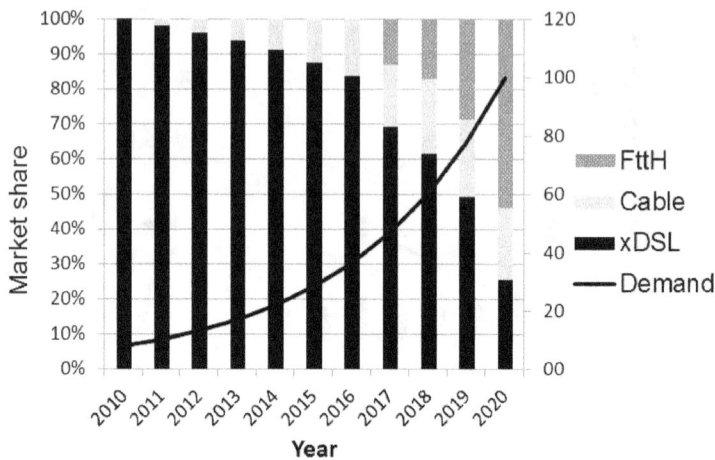

Figure 3.34: Market share per year.

intervals. A larger network usually is more cost-efficient per user. In our example, our 10.000 Euro a year network can sustain up to 200 customers. After that, we need to upgrade to a 20,000 Euro a year network, which can sustain up to 1000 customers. If we have only 201 customers on this network, network cost will be near 100 Euro a year, but if we can reach a 1,000 customers, network cost will have dropped to only 20 Euro a year per customer. These declines in cost per customer are often referred to as scaling. For the model we need to know the scaling parameters and the scaling points.

3.4.3 Case 1: migration path

3.4.3.1 Description of the case

If an operator has as starting position the Full Copper topology in a certain area, he has to decide on the next step: bring the fibre connection all the way to the customers or use an intermediate step, where he brings the fibre closer to the customer, e.g. FttCab. To make this decision he has to look at the pros and cons of all the options. For example, the implementation of FttCab can be much faster than Full FttH, as it requires less digging, the last part of the connection from the street to the access node in the house does not have to be installed, and it meets the growing bandwidth demand for now and the near future. If, in future, this demand exceeds the supplied bandwidth, the remaining part to the residence can be connected with Full Fibre or using Hybrid Fibre as extra intermediate step. If the demand does not exceed the supplied bandwidth, for example it reaches some level of saturation, no further migration is needed, saving a lot of investments. However, when Full FttH is the expected final solution, using intermediate steps would incur investment and installation costs that might be lost and not reused.

In this case we present a gradual topology migration path from Full Copper to Full FttH, where we look at financial impact in relation the FttH-direct opportunity. In the

3.4 Economic impact

migration path we want to reuse tubes, cables and fibres or prepare them as much as possible. Preparing means that it is possible, for example, to put extra tubes in the ground when rolling out FttCab, that you will need in the case the full FttH step is made. This saves an extra digging activity later on. Of course not all equipment can be reused and not all pre-investments will be economical, but we will show that the postponement of huge investments will recover a part (or all) of these extra costs.

Three topology migration steps are:

1. From Full Copper to FttCab - see Sections 2.2.1-2.2.3.

2. From FttCab to Hybrid FttH - see Section 3.2.

3. From Hybrid FttH to Full FttH - see Section 3.3.

If an operator follows the gradual topology migration path he better takes the possible next steps into account when planning the first step, FttCab. To do this he can take the following steps: **Step 1: Dimensioning the circuit.** When the operator makes the step to FttCab, the maximum number of customers per FttCab circuit, say K, is related to the number of customers connected to a Full FttH-PoP, call that X. This Full FttH-PoP (PoP in the remaining of the chapter) will be necessary in the case Full FttH is rolled out and the cabinets are not big enough to handle the active equipment. The parameters X and K together determine the number of PoPs per circuit:

	clients/PoP	clients/ring	PoPs/ring
1	$0 < X < 0.5K$	$\lfloor K/X \rfloor * X$	$\lfloor K/X \rfloor$
2	$0.5K < X < K$	X	1
3	$X > K$	K	< 1

Suppose $X = 500$ and $K = 2900$: Here situation 1 is applicable: there will be $\lfloor 2900/500 \rfloor \cdot 500 = 2500$ customers allowed on the circuit and five PoPs. See Figure 3.35.

Suppose $X = 1500$ and $K = 2900$: Here situation 2 is applicable: there will be 1500 customers allowed on the circuit and one PoP.

Suppose $X = 6000$ and $K = 2900$: There will be 2900 customers admitted to the circuit. The CO location will (if possible) service numerous circuits and (possibly) contain numerous PoPs. See Figure 3.36.

Step 2: Setting up FttCab architecture. The architecture of the FttCab implementation can now be determined using the method described in Sections 2.2.1-2.2.3. When clustering the cabinets, a precondition should be the number of customers (maximum) on the fibre circuit, and therefore the cluster, of step 1. When creating the circuit it should be taken into consideration that the circuit will go through the cabinets and through the determined number of Full FttH POP location(s) from step 1. Here also extra ducts should be placed for latter Full (PtP) FttH delivery.

Step 3: Setting up Hybrid FttH architecture. Here the duct structure has to be continued to the Hybrid FttH node. Not only the (possible) one fibre is needed to connect the Hybrid FttH node, but be prepared for the Full (PtP) Fibre roll-out.

Step 4: Transition to Full FttH. When the time has come to transmigrate to Full FttH, two situations might be possible:

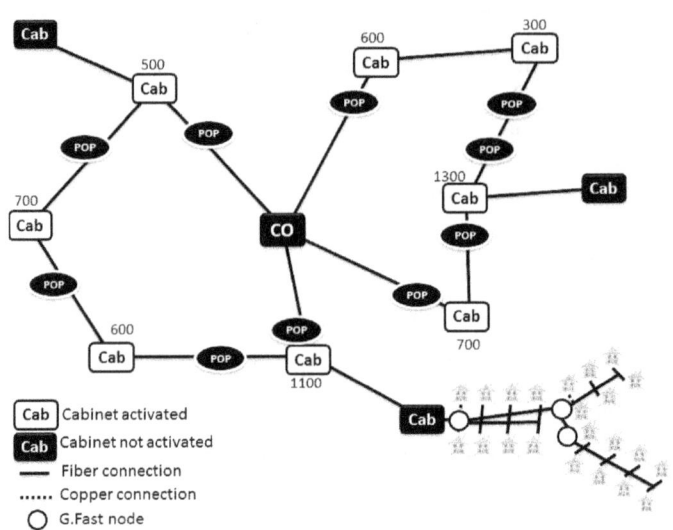

Figure 3.35: Example with $X = 500$, $K = 2900$.

1. More than one Full FttH-PoP per ring: Every residence receives the fibre optic connection to the original cabinet. Households (or cabinets) are allocated to Full FttH POP in such a manner that the total distance is minimized (within capacity limitations).

2. One or less Full FttH-PoP per circuit: Households receive fibre optic connection according to original CO location.

3.4.3.2 Numerical results

The area we look at has a density of 6000 connections per square kilometre, a city centre. These assumptions are representative for the Dutch case, but also for other countries where the last 20-200 meters are constructed by underground cables. We take an area with 2304 connections, which is 1 full CO and a G.Fast node capacity of 48 connections.

The first topology migration path under consideration is the presented gradual topology migration path, from Full Copper (FC) to Full FttH (FF), using FttCab (FCab) and Hybrid FttH (HF) as intermediate steps. In Tables 3.14 and 3.15, the costs of the three steps in the topology migration path are shown. The categories are those of the previous section.

This includes 2% inflation and adds up to a total of € 1,280 per connection when the total topology migration path is followed. Note that bringing only FttCab and Hybrid FttH to the customers is relatively cheap, only € 292 for bringing already a high bandwidth. The roll-out of Full FttH directly migration will lead to the following calculated costs: This is a total cost of € 968 per connection. Around 30% cheaper. But now will use the discounted cash flow (DCF) method to compare the two outcomes. The used weighted cost of capital (WACC) for fixed telecom operators comes from [104] and is 7.38%. If

3.4 Economic impact

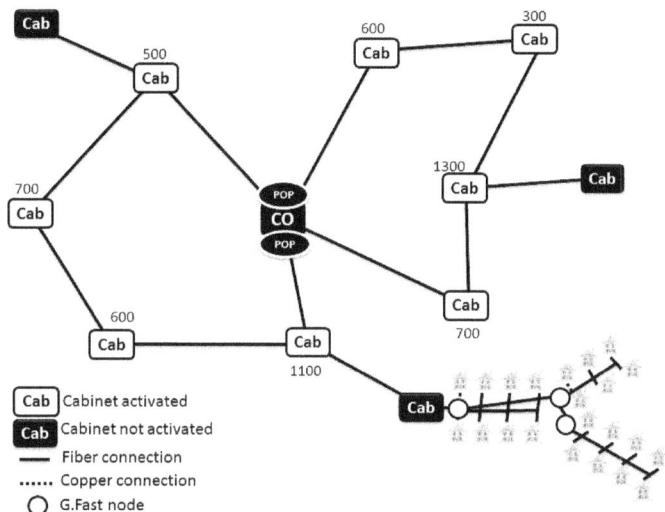

Figure 3.36: Example with $X = 6000$, $K = 2900$.

Category	FC to FCab	FCab to HF	HF to FF
(1)	€ 87,930	€ 3,000	
(2)	€ 60,000	€ 1,500	
(3)		€ 98,545	€ 24,000
(4)		€ 76,800	€ 12,000
(5)			€ 1,434,359
(6)	€ 115,200	€ 230,400	€ 806,400
Total	€ 263,130	€ 410,245	€ 2,276,759
Per connection	€ 114	€ 178	€ 988

Table 3.14: Costs of the three migration paths.

we assume that the investment for FttCab has to be made next year, the migration to Hybrid FttH will be in five years (on average) and the migration to FttH will be in 10 years (on average) and compare this to a Full FttH roll out next year (again on average) the cost comparison is totally different, regarding to the discounted cash flow (DCF) as shown in Table 3.16. Now the migration path is cheaper, € 836 against € 968 in net present value.

In Figure 3.37 we see the difference in discounted costs for different values of density of connection, ceteris paribus, and in Figure 3.38 the maximum copper length with different density of connections, ceteris paribus. Both are calculated for both a 16-port G.Fast node and a 48-port G.Fast node. This are the two options that are under consideration of the manufacturing parties. We can conclude that, with the chosen parameters the migration path is cheaper in DCF than the direct roll-out of Full FttH for the 48-port. In the case of the 16-port the break-even point is touched at 6000 connections per square kilometre.

Category	Full FttH
CO to premises	€ 1,424,229
Premises	€ 806,400
Total	€ 2,230,629
Per connection	€ 968

Table 3.15: Total costs of the three migration paths.

Migration	FC to FCab	FCab to HF	Hf to FF	FC to FF	Total
Gradual	€ 114	€ 137	€ 585		€ 836
Full FttH				€ 968	€ 968

Table 3.16: DCF of the two paths.

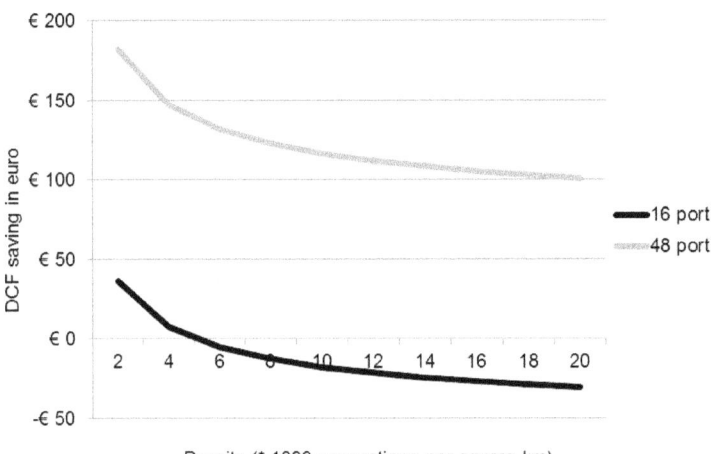

Figure 3.37: Saving DCF migration path.

3.4 Economic impact

Some observations:

- With a star structure the distances will be shorter so this can serve more nodes. The Dutch situation (see Figure 3.2) is heavily branched which has both characteristics of snake and star networks. You can find points in this network from where it looks like a star structure further on in the direction of the customer.

- Bonding, combining multiple wire pairs to increase available bandwidth, will reduce the capacity (in connections) of the node.

- The maximum copper distance used for VDSL varies between countries and studies. In the Netherlands a maximum of 1000 meters is used, whereas in practice in the cities a large part is within 200 meters. The study presented in [155], as discussed earlier, adopt a network structure with a two layer cabinet solution with Branch Micro Switches (BMS) and Lead Micro Switches (LMS) where the BMS is connected with the CO with two paths and the LMS is connected to two BMSs. The users are connected in a star with one LMS with a typical distance of 100-300 meter. [129] see a typical maximum VDSL distance of 400 meter. This all indicates that in several countries a roll-out with Hybrid FttH nodes at the current cabinet is possible in a big part of the cases.

- We compared the two topology migration paths until Full FttH. However, the gradual topology migration path has the option that on of the intermediate steps will be the final solution if a level of saturation in bandwidth demand is reached. This leads to lower expected costs or some real option value (see for example [4]), that was not taken into account in our approach. This could justify the 16-port case. Next to this, a fast roll-out of FttCab could save the market share of the operator, see for example the second case in the next section.

- From technical point of view, the 48-port modem is not just a combination of 3 16-port modem. To serve a bigger group connections that share a cable, they should be served from the same modem that makes complex calculations to reduce the crosstalk effects.

- A point of concern is that in case of a 48-port G.Fast modem not all connections are within 200 meters over copper from the Hybrid FttH node (see Figure 3.38), which is, about, a bound for the high bandwidth using G.Fast, see [19]. In a case study of Amsterdam, see Section 3.2, based on real distances, locations of cabinets and copper cables we saw an average utilisation of the G.Fast node of 38 ports to serve 99% of the connections within 200 metres.

3.4.4 Case 2: total financial impact

3.4.4.1 Description of the case

The model works with scenarios. A scenario consists of percentages describing the coverage per technology per year. The percentages refer to the availability of the most

118   FttH planning and economic impact

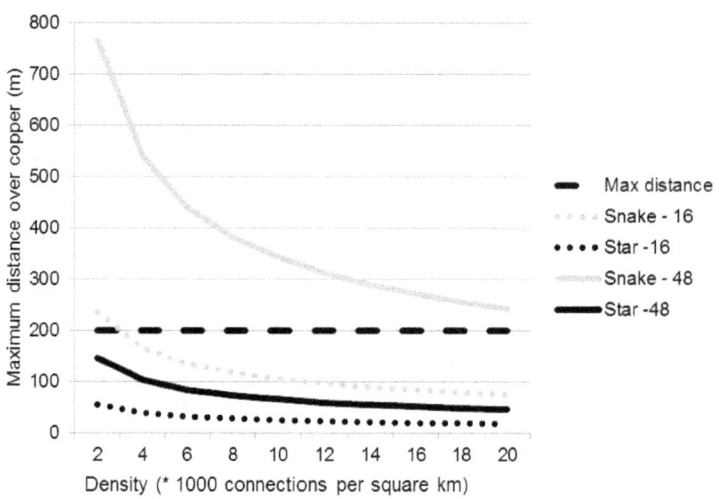

Figure 3.38: Maximum copper length snake structure.

advanced technology, meaning that a household that is both covered by FttH and xDSL will be counted as a FttH household. The sum of the coverage percentages per year should therefore always be 100%. To determine which technology is the most advanced, the following technology ranking is used: xDSL < Hybrid FttH < Full FttH. Next to this a near 100% cable TV coverage is assumed.

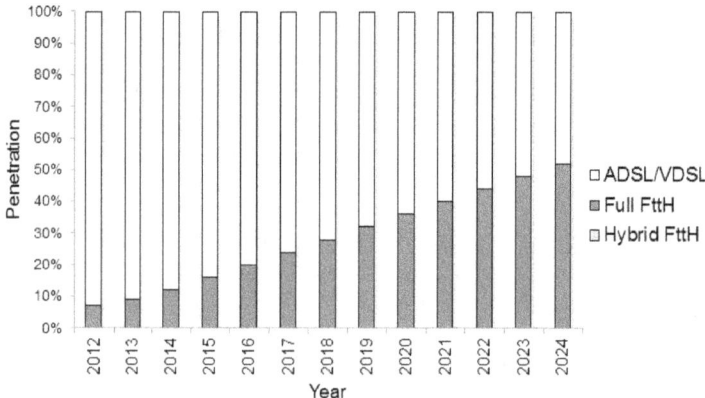

Figure 3.39: Penetration of technology with FttH growth as base scenario.

For the analysis we take a base scenario in which Full FttH grows with 4% per year and there is no Hybrid FttH, see Figure 3.39. We think that this is a conservative scenario. To realize this scenario we assume:

- A certain adoption rate: the number of households that is willing to accept a high

3.4 Economic impact

bandwidth connection (e.g. in relation to the price).

- A budget: the amount of money available for investment by a telecom operator.
- A certain amount of available labour: how many connections of each technology can be realized per year.

We argue that these are the three main constraints of an operator when migration to a FttH network. To compare this base scenario to a scenario where Hybrid FttH will be used, we look at these three constraints:

1. Keep the adoption rate constant, shared between Full en Hybrid FttH.
2. Keep the amount of available labour constant, assuming six Hybrid FttH connections cost as much construction capacity as one Full FttH connection.
3. Keep the amount of available budget constant.

We assume that every operator has one dominant constraint. Now we are interested in the question: given the dominant constraint, is introducing Hybrid FttH business wise (revenues, earnings, market share) interesting for an operator? The calculations are not based on an existing operator, but is very similar to the Dutch case. For more analyses and results of sub-models see [20]. We will look at the three cases in the following sections.

3.4.4.2 Effect of introducing Hybrid FttH under constant adoption rate

In the first case we consider the demand for 4GBB connections limited, a part of this demand can be fulfilled by a Hybrid solution. Starting at 2015, 40% of the new connections is realized by Hybrid FttH[12]. Under this constraint we compare the base scenario with the scenario that Hybrid FttH is introduced. The same number of 4GBB-connections are realized, but in different technologies. This is shown in Figure 3.40.

Result of this comparison, see Figure 3.41, is that introduction of Hybrid FttH leads to:

- Lower cost,
- Lower investment,
- Slightly higher revenues,
- Higher earnings,
- Almost same market share,
- DNPV = € 760 million.

In this case the operator is able to fulfil the demand from the market with lower costs, leading to higher earnings.

---

[12]This number is an estimation made in the project of the percentage of houses in the Netherlands that are suitable for G.Fast.

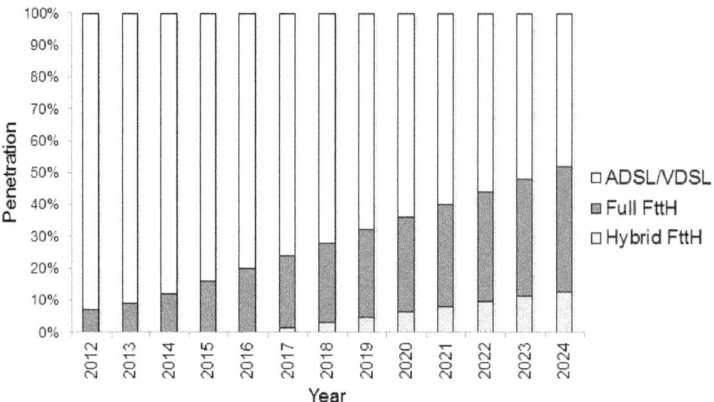

Figure 3.40: Penetration of technology in case of constant adoption rate.

3.4.4.3 Effect of introducing Hybrid FttH under constant building capacity

Here we look at two scenarios where the building capacity is kept the same. Under this constraint it is possible in the second scenario to realize more Hybrid FttH connections with the same labour, we assume a factor 6. This means that in the effort needed for one Full FttH connection the operator can realize six Hybrid FttH connection. This factor is based on the difference on digging distances. This leads to more realized 4GBB-connections. This is shown in Figure 3.42.

Result of this comparison, see Figure 3.43, is that introduction of Hybrid FttH leads to:

- Slightly higher cost,
- Almost same investment,
- Higher revenues,
- Higher earnings,
- Bottoming-out of market share,
- DNPV = € 2,067 million.

In this case the operator is able to realize more connections for almost the same investment. This leads to a stop in the declining of the market share, what leads to higher earnings in later years.

3.4.4.4 Effect of introducing Hybrid FttH under constant investment

Here the money available for investment is kept the same. It is possible to realize more Hybrid FttH connections with the same amount of money, while these connections are

3.4 Economic impact

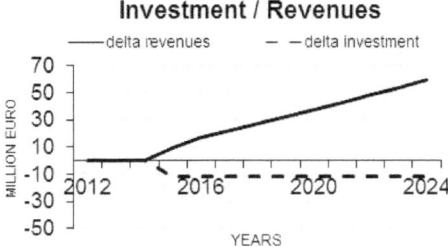

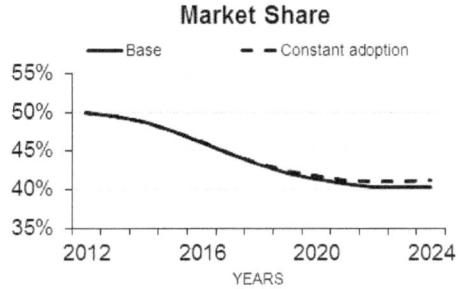

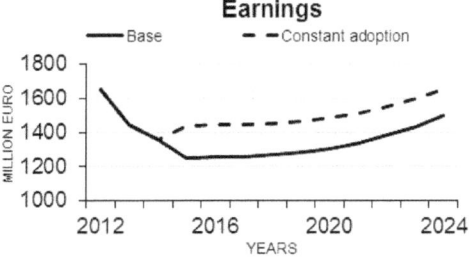

Figure 3.41: Results constant adoption rate in million Euro.

cheaper to realize. This leads to more realized 4GBB-connections than in the base scenario. This is shown in Figure 3.44.

Result of this comparison, see Figure 3.45, is that introduction of Hybrid FttH leads to:

- Slightly higher cost,
- Same investment,
- Higher revenues,
- Higher earnings,
- Bottom under market share,

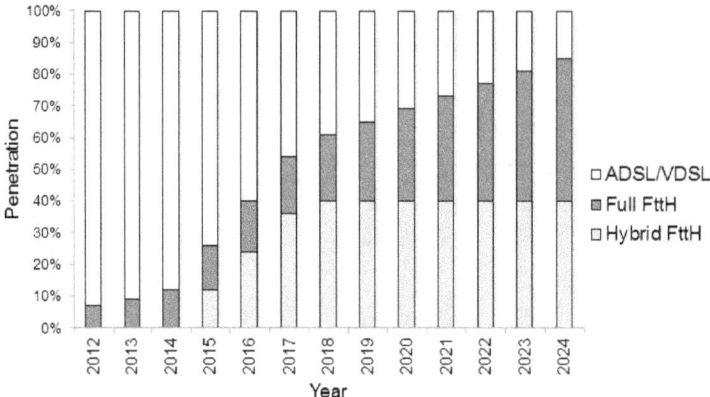

Figure 3.42: Penetration of technology in case of constant building capacity.

- DNPV = € 729 million.

The extra investment in scenario 2 is very low. This leads to the situation that the results of this scenario is almost the same as scenario 2.

3.4.4.5 Conclusions

In all three cases the Hybrid FttH connections are the final solution for the subscriber; there is no need for further migration towards Full FttH. This gives in all cases a cost saving and increase of earnings. The real benefit is in the increase of the market share, which indicates the viability of the operator. All three cases give a increase in market share, but in the case of constant building capacity the increase is significant.

In the base scenario (see Figure 3.39) was chosen for a FttH connections growth of 4% per year. In Figure 3.46 the sensitivity of results to this percentage is shown. Here the 2024-market share is shown under four growth rates.

3.5 Summary

First the planning of the FttCurb variant using G.Fast as technology was studied, where the fibre is brought to a place in the street, also known as Fibre to the Curb. To realize FttCurb using G.Fast a next step in bringing fibre to the houses is needed. Here a new node is realized within 200 meter of each house connected. We assumed that a branching point in the existing copper connections is chosen to place the new active equipment. The new nodes have to be connected by a fibre connection. A framework that is based on three main choices before designing the network was presented. If these three choices all have two options, we end up with eight possible planning options and 6 main mathematical challenges, which have been all elaborated in this chapter, showing the mathematical approach for all of these options. For one of the options the results of a real life case was presented, the planning of FttCurb in Amsterdam and The Hague.

3.5 Summary

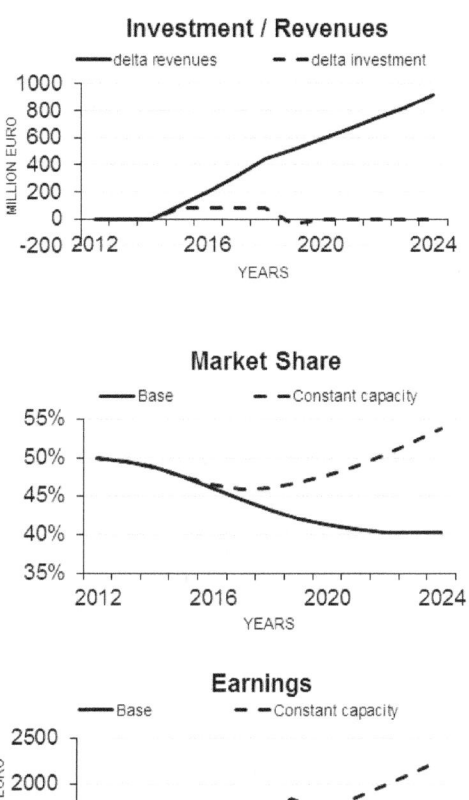

Figure 3.43: Results constant building capacity in million Euro.

The roll-out of a fibre network, although necessary, generates both high direct costs and social costs. First a model was presented to optimize the location of the concentration point and expanded the standard model with some details to make the answer more useful in practice. The problem of finding an optimal method of laying this network was translated into a mathematical problem in a novel way: we know already where the fibre segments has to be, but how to connect them by a coherent network? This was described by a Steiner Tree Problem with costs both on nodes and edges. To solve this an algorithm was proposed that translates the underlying graph with costs on nodes and edges to a graph with only costs on edges without Steiner Points. Then we solved the remaining Minimum Spanning Tree and translated it back to the original problem. Also a method was introduced to incorporate smart co-laying opportunities which reduces the social costs, in the planning by modelling it as a Steiner Tree Problem with timing

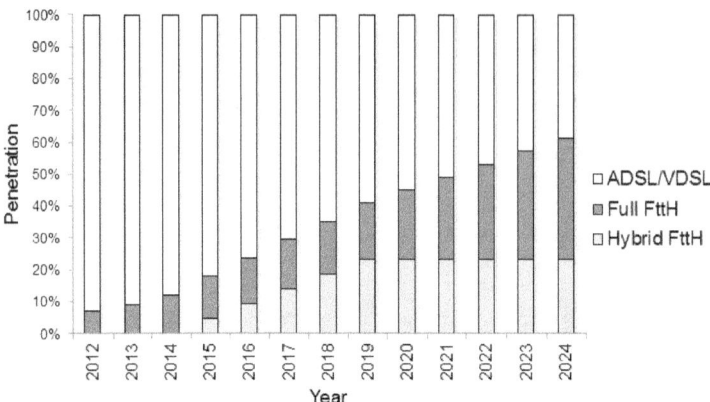

Figure 3.44: Penetration of technology in case of constant investment.

benefits. The methods have been illustrated with small examples. These heuristics have been implemented into a tool that can support both municipalities and entrepreneurs in their choice of the optimal method for laying the fibreglass infrastructure.

Next the techno-economic model we designed and built has been shown. This model can calculate the effect of using different technologies in access networks on market share, revenues, costs and earnings. We introduced a new method to calculate churn rates. Then we looked at two cases to show the working of the model. We first looked at the economics and planning issues of a full migration path from ADSL to FttH, using FttC and Hybrid FttH as intermediate steps. We outlined a possible migration path that can be used in practice. We discussed some planning issues that arise in each migration step and the precautionary measures that have to be taken for steps in the future. We calculated the differences between the proposed migration path and the one-step Full FttH roll-out and showed that, under our assumptions, which are representative for the Dutch case, but also for other countries where the last 20-200 meters are constructed by underground cables, the migration path is economically feasible. For an operator this is important information: bringing Hybrid FttH is a relatively cheap option to deliver high bandwidths quickly. If, however, this solution will be insufficient in the future, the postponement of the investment of FttH gives a cost saving that is big enough to compensate for the extra costs of the full migration path. In the second case we argued that in rolling out FttH, operators are restricted by either budget, labour or customer demand (adoption rate). In all three the cases using hybrid FttH gives slightly lower costs but much higher earnings now and in the future. In the case that labour is the bottleneck of the operator, there is an important gain in market share if the operator starts rolling out Hybrid FttH. For these operators this could be the way to survive the battle against the cable operators: a hybrid FttH solution can offer more clients the demanded bandwidth in the same time. This is very important when a telecom operator is faced with the competition of a cable TV operator. This should be the real reason for a telecom operator to start with a Hybrid FttH solution using G.Fast and not a possible short term cost saving. Also the model could be used to weight the different technology

3.5 Summary

Figure 3.45: Results constant investment in million Euro.

options in a case by case approach. If an operator wants to bring 4GBB connection to a certain area, what would be the most economical way, considering the number of apartment buildings, the building density, the available number of copper pairs etcetera.

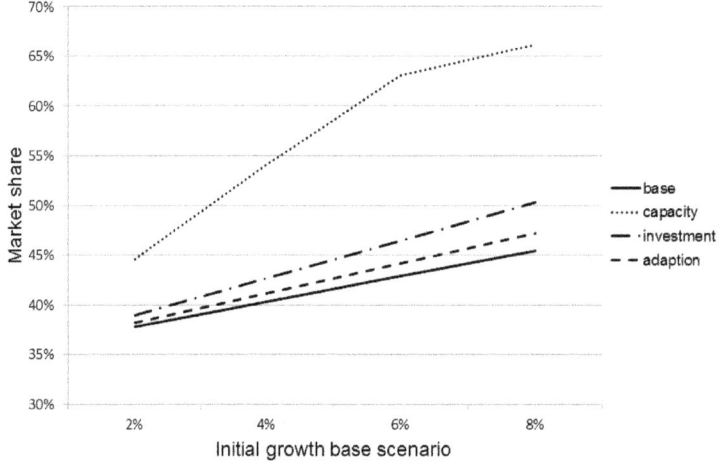

Figure 3.46: Market share sensitivity analysis.

4 Electricity network planning

Recent power market developments and the increasing penetration of renewable energy such as wind power, account for growing uncertainties both in demand and in supply of power. This affects not only the issues of power quality and voltage regulation in short term power system operation; it also has consequences for general grid performance measures such as congestion risk and transmission losses. In this chapter a tactical planning view on placing of distributed generation to minimize transmissions losses is presented. This is done at two levels: (1) what is the optimal allocation within a district, and (2) what is the optimal placing of wind turbines over de country. This chapter is based on [34, 35, 93].

4.1 Introduction

This chapter is started with some background on the problem and the available literature.

4.1.1 Tactical generation planning

As was sketched in the introduction of this thesis, the producers of electricity and the managers of the networks have to deal with increasing demand and an increase in highly variable production by local generators such as solar cells and wind turbines. The challenge will be to optimally integrate the increasing number of (small) generation units in an electricity system that up to now has been very centralized, integrated and planned (see e.g. [115]). Since most DGs rely on exploitation of natural sources of energy they exhibit high fluctuations in production over time. This means that electricity generated by DGs will probably not match load demand and can cause over- or underproduction of electricity. The current technological solution to solve possible transport problems is to reinforce the existing grid as stated by Gitizahed [51]. This is very expensive and is avoided as much as possible. Another solution, still in development, is to make the grid smarter by controlling fluctuations in production and consumption, using smart grids (see Kok [80, 81]) and production planning (see for example Kopanos et al. [84]). Next to this, many trials have been done by managing the behaviour of customers, like in Faruqui et al. [45].

Before grid reinforcements are effectuated and smart grid are created that can perform sophisticated load balancing network owners should have some idea about the global demand and supply characteristics and perform *tactical planning*: try to find an optimal mix, in size and nature, and placement of DGs while keeping network capacities in mind. All demand has to be satisfied by power plants and/or DGs. Each DG type has a different production pattern depending on e.g. the sun, wind or heat demand, that may complement each other. When there is sunshine (or wind) all PV solar panels (or micro wind turbines) in the district will generate electricity at the same time, leading to a sudden increase in generated electricity.

In this chapter two main problems are considered. In the first problem we consider a single district; if all houses in a district have the same DG type, then it will probably be less efficient compared to a district with an optimal mix of DGs. Hence, finding such an optimal mix can provide a better understanding of the effects of DGs on energy loss in the current power grid. The search for the optimal penetration level[1] of DGs is graphically depicted by Scheepers and Wals [132] in Figure 4.1. This graph is based on a study by Mendez et al. [108] where they modelled several scenarios of DG penetration levels, DG dispersions and DG mixes. The idea is that because the distance between feeder and load is reduced, the transportation losses decrease. But as Figure 4.1 shows, this only applies to a certain DG penetration level. Once the optimal penetration level of DGs is reached the losses increase again. Higher DG penetration levels lead to situations where local production exceeds local consumption. This overproduction has to be converted into higher voltage levels and transported further away. Converting electricity from one voltage level to another creates loss and transporting electricity also creates loss, which thus means that overproduction will ultimately create more loss. So if too many houses start generating electricity, the increase in penetration levels of DGs becomes more of a disadvantage than an advantage. This graph gives a good illustration of what we are looking for.

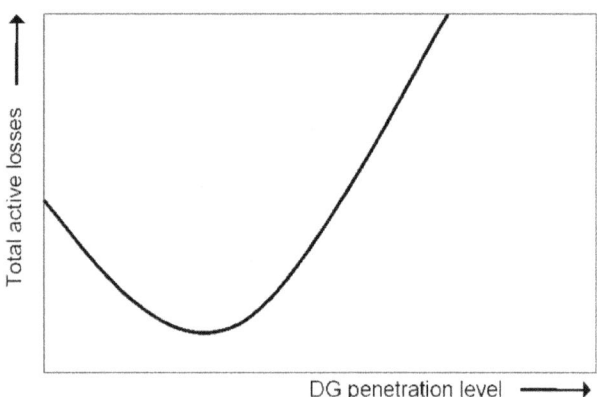

Figure 4.1: Variation of distribution losses due to DG penetration, [132].

To the best of our knowledge we are the first developing a model for tactical DG planning, for size and nature, within a district, minimizing total transportation loss and avoiding overload. Kools [83] mentions three important issues when evaluating the

---
[1] Penetration level = ratio of capacity factor times total DG power installed and the peak power demanded on the feeder.

4.1 Introduction

effect of DGs: (1) the capacity of the generators, (2) the type of the generators and (3) the location of the generators. These issues are dealt with in various combinations in literature. We focus on the combination (1) and (2) what has not been done so far. In Section 4.2 the mathematical representation of this problem is presented. Next the used data and the limitations of the retrieved data is introduced and the solution method of the chosen optimization model is explained. Finally the results of the selected case is presented, together with conclusions and some thoughts on the implementing issues of the optimal mix of DGs.

The second problem in this chapter considers a normal load situation of a High Voltage (HV) network and tries to find the optimal locations to build a given number of new wind turbines, in order to minimize the expected energy transportation losses. Here two trade-offs have to be managed:

1. Local or central placement: loss percentage for transportation of a certain amount of electricity is proportional to travelled distance. If all demand and production of power on the grid were deterministic and time-invariant (which is a purely theoretical case), transportation losses would be minimized by placing all generators as close to demand points as possible. But when demand and production are time-varying or stochastic, sometimes local production is higher than local demand. Local excess power in peripheral parts of the grid has to be transported further than if it were produced at some central point. So local placement of generators is advantageous when local demand is larger than local production, but disadvantageous when local demand is smaller than local production. In that last case central placement would be better. Therefore loss-minimizing placement of stochastic generators may be regarded as a kind of trade-off between the advantages of local and central placement.

2. Spreading or concentration: the correlation between the stochastic production of wind turbines reduces when they are placed further apart. So a certain spreading of the wind turbines will result in smaller variance of total wind power production, and therefore may reduce the disadvantage of its stochastic nature. However, spreading of the wind turbines may have to mean that some are placed in regions with less wind, which in itself is undesirable. Also in this respect a trade-off has to be made.

In Section 4.3 a three step procedure to solve this problem is presented. The first step is to apply the well-known DC power flow model to the network structure and the load data. The second step is to specify a set of scenarios describing the wind power production behaviour of the wind turbines to be connected to the grid. Each scenario is described by a different wind power output and its proper probability, all based on real wind data. The third step is to optimize expected power losses over all different possibilities of wind turbine placements using a model for stochastic optimization, solved quickly by a novel, very simple, but well justified heuristic.

4.1.2 Literature review

There is a wide range of different types of energy models with different approaches and objectives. The review papers of Jebaraj and Iniyan [71], Hiremath et al. [63] and Tan et al. [141] give good overviews of energy models presented in the literature. In [71] an overview is given of energy models that have been emerging over the last few years. The following types of energy models are discussed: energy planning models, energy supply-demand models, forecasting models, optimization models, energy models based on neural networks and emission reduction models. Tan et al. [141] give a thorough overview of solution methods for deterministic models for optimally distributed renewable generation planning.

From the energy planning models two articles stand out: [89] on special programming models and [12] on the pay-off matrix technique. In [89] elementary spatial programming are discussed such as Quadratic Programming (QP), Mixed Integer Linear Programming (MILP) and linear complementarity programming models. These models are used to optimize energy production, transportation, distribution and utilization with respect to cost.

In [12] an approach is presented by Belyaev to the solution of decision problems using a pay-off matrix. Stochastic variables are used to represent uncertainty for data that are not precisely known. These are modelled for several scenarios and are put into a pay-off matrix. Using this pay-off matrix a decision is made.

In the articles on optimization models more often than not the objective is to minimize costs. See for example [38], [131], [139], [98], [26] and [63]. Those papers minimize cost or maximize GNP/energy ratio and try to find an optimal mix of energy sources. Even though there may be some resemblances, there are substantial differences between these papers and our problem. Firstly our objective is to minimize energy loss and thus means that we have a different (quadratic) objective function. The second difference is the type of variables that is used in the optimization model. Energy sources can be petroleum or wood, and DGs can be wind turbines or solar panels. To model DGs one uses binary variables, while when modelling energy sources the decision variables are non-negative real numbers. So these models cannot be used to solve our problems.

When searching the scientific literature for minimization of power losses in a network, it becomes clear that there exists a long history of research on this topic, and that it is now vivid as ever. The major part of the research papers is on subjects such as the optimal placement of capacitors on a distribution line, or the optimal configuration of sectionalizing switches in a near-radial distribution system, for example Ababei and Kavasseri [1], Baran and Wu [11] and Rao and Narasimham [127]. Newer is the interest in using Distributed Generation to minimize transportation losses, see for example Wang and Nehrir [151], Le et al. [92], and Quezada et al. [126]. There are three other papers that are more related to our first problem. Singh [135] minimizes energy loss for optimal sizing and placement of distributed generation using a genetic algorithm. This paper discusses a simulation approach for the optimal sizing and placement of a DG for a minimum annual energy loss with time varying load model. It is made sure that the voltage levels are within the acceptable range and that the line flows are within limits. This paper focuses on the quality aspects of electricity, so voltage levels and reactive power are also considered. This means that the model in [135] is too complex for our purpose. In [103] the authors presents a multi period optimization model for a micro grid, aimed at maximizing its benefit, i.e. revenues-costs. The optimization model includes the use of DGs relying on wind and solar, an electrochemical storage and interruptible load. DGs are incorporated into the low voltage grid where both technical and economic aspects are considered. The obtained problem is a Mixed Integer Non-Linear Programming (MINLP) problem and is solved using a genetic algorithm. Even though this model minimizes cost, we obtain a model with similar structure due to the fact that it also tries to find an optimal mix of DGs. This similarity is especially present in the constraints of the storage system and DGs. Mashhour and Moghaddas-Tafreshi have gone one step further and allow the DGs and storage system to be controllable. This creates a system resembling the smart grid. In [44] a method is presented for locating and sizing of DGs, with respect to mainly voltage stability and a reduction of network losses. They use a static approach and optimize using dynamic programming.

For the second problem we are interested in wind power modelling. While the field of renewable energy is constantly growing, and wind power is one of its largest contributors a lot of research is done into wind power modelling under two main themes: firstly, *steady state analysis*, which is mainly used in long term planning, power grid design and generation expansion studies; and secondly, dynamic analysis (or, more commonly: *forecasting*). This is used for short term prediction of future wind power generation, in order to optimize intra day market biddings and short term power system control. Especially because of the reactive power behaviour of wind turbines, voltage stability in the system may depend crucially on their correct operation. Examples of earlier work are Orfanos et al. [112], Hu et al. [66], Ivanova [69], Feijoo et al. [46] and Papaefthymiou and Kloeckl [114].

More specified to our second problem, namely the inclusion of wind turbines in an existing grid, is found in Carpinelli et al. [21], who addresses the question of 'distribution system planning in the face of a worldwide growth of DG penetration'. The aim of their article is to present a 'proper tool, able to find the siting and sizing of DG units which minimize generalized cost' as a help for the distribution system planner. They designed a three-step algorithm to find the optimal locations and sizes of several wind turbines, to be built within the area of some Medium Voltage (MV) distribution grid. The first step consists of an implementation of the method of [46] mentioned above; the second step is a genetic algorithm which steadily improves the intermediate solution. The objective function (generalized cost) contains several terms, such as location dependent building costs, but also the cost of power losses. The third step is a method of determining the robustness of a solution under uncertain future conditions. Lastly, they apply the method to a particular Italian MV grid.

Chen [27] wrote a PhD thesis on stochastic network design. He takes his applications from design in logistic networks and power systems. The subject of chapter 4 is Transmission and Generation Expansion Planning (TGEP) for wind farms in power systems. He shows that for wind energy, the expansion planning of generation capacity and of transmission capacity should be considered simultaneously. He shows that however very little work has been done on the simultaneous TGEP problem, mostly generation and transmission expansion are regarded independently, or else a deterministic TGEP model is used. But wind power requires a stochastic model.

Next he presents a two stage stochastic programming formulation of the integrated TGEP problem. The first stage decisions represent the design decisions: where to build generators, and of which type; and what transmission capacity is needed, and of which type. Scenarios reflect all randomness in availability of generator capacities and transmission line capacities and in the demand at the demand nodes. System reliability is promoted by putting a penalty on high expected loss-of-load in the objective function. Other terms in the objective are the installation costs of the expansions, and the expected operation costs. Operating costs of a generator for a certain scenario is a linear function of the generation. Summing over all generators and scenarios yields the expected operating cost.

A total different approach is found in Kramer et al. [87]. This paper presents an approach to calculate the optimal location of renewable energy generation. Potential areas for the expansion of the considered generation technologies (wind onshore and offshore, photovoltaics and concentrated solar power) are calculated in a land use analysis. The corresponding renewable energy supply (RES) is determined via a meteorological analysis. Eventually optimal RES investment decisions are determined. Wind energy should mainly be expanded on coastlines but higher capacities need also efficient interior locations.

Although our research is related to [27] there are some differences in the approach. We go into much more detail of the power grid and use a different model for it. Next to that, as our model is very complex and simplification does not help enough, we present a simple heuristic to generate solutions. Our approach, looking at transportation losses and taking into approach the correlation in wind power production lead to conflicting conclusions.

4.2 Optimal mix of distributed generators

Distributed generation can increase efficiency in the grid by reducing the distance between generators and consumers of electricity. However, adding too much correlated generators will cause overproduction and thus losses. The objective of this chapter is to introduce tactical planning, by finding an optimal mix of distributed generators in a district such that energy loss is minimized and overload is avoided. The effect of using future electronic devices, such as electric vehicles, is also studied. To find an optimal mix of distributed generators a mixed integer quadratic programming model is defined and a case study is presented.

4.2.1 Problem definition

In this section a mathematical optimization model is presented to find an optimal mix of DGs such that energy loss is minimized. The focus is on a district consisting of only houses, where each house can generate a part of their demand using DGs. When there is overproduction, it is supplied back to the grid. Also a storage system in the district is included such that overproduction can be stored. The optimization model needs to make sure that demand is always satisfied and that overload is avoided as much as possible.

4.2 Optimal mix of distributed generators

First the overall interpretation of the model is discussed, such as how variables are defined, what the assumptions are, how energy loss is estimated and how overload is avoided. Then the different parts of the mathematical model are presented.

4.2.1.1 Some initial assumptions

An electricity supply system comprises of three main components: power generation, transmission and distribution. In Figure 4.2 a simple diagram of the Dutch power grid is shown. The High Voltage grid will be referred to as the HV-grid, the Middle Voltage grid as the MV-grid and the Low Voltage grid as the LV-grid.

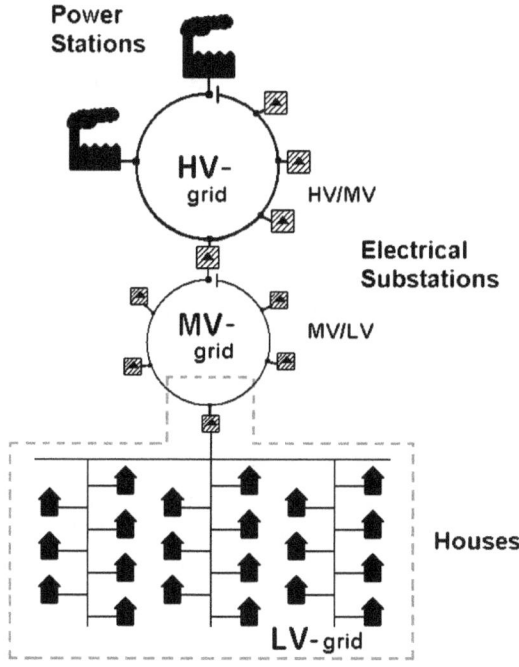

Figure 4.2: A diagram of the Dutch electricity grid.

The HV-grid is comprised of large power plants connected to multiple HV/MV substations. Electricity is transmitted through high voltage cables (110, 150, 220 and 380 kV) to reduce energy lost in transmission. The MV-grid connects a HV/MV substation to several MV/LV substations. In HV/MV substations a transformer converts high voltage power into middle voltage and in MV/LV substations middle voltage is transformed into low voltage. MV-grids transmit electricity with power between 10 and 20 kV. Houses are connected in series to a single linear cable, also referred to as the trunk. Multiple trunks are then connected again to a single linear cable, which is connected to an MV/LV substation. In LV-grids electricity is transmitted at 0.4 kV.

Transmission refers to the transportation of high voltage electricity over long distances (HV- and partly MV-grids) and is managed by the Transmission System Operator. The distribution of electricity has to do with the transportation to consumers in the grid, i.e. LV-grids and partly MV-grids. The distribution grids are managed by the Distribution System Operators. Most power stations in the Netherlands are natural gas power plants but there are also some coal-fired power plants, biomass power plants and one nuclear power plant. There are a few offshore wind farms and a few small hydro-electric power stations.

In this section, we focus on a district connected to a MV/LV substation. This is illustrated by the selection in Figure 4.2. The reason that we choose to model only houses is that there are too many different types of industries and businesses, from mining to offices and swimming pools. Each have quite a different demand pattern which makes it difficult to model. All houses can generate a part of their demand using decentralized power generation; any overproduction is sold back to the grid. So there will be a bi-directional flow of electricity in the grid. We also include storage systems in the model to help capture overproduction and, if possible, reduce energy loss. Whether it will actually reduce energy loss depends on how efficient the storage technologies are.

Since the focus is on one district connected to an MV/LV transformer, we do not model power plants, neither other districts' production and demand. Therefore we need to make some assumptions on their behaviour. To make sure that there is enough electricity generated to satisfy the whole district's demand, power plants are assumed to supply the remaining demand of electricity. And because we do not know how much electricity is generated in other districts, we assume that the district only imports electricity from power plants. Furthermore, we assume that electricity exported to outside the district can always be consumed by another district. This way we avoid the issue of whether other districts can consume the overproduced electricity or if it will be wasted energy. To summarize, power plants always make sure that demands are satisfied and the other districts always make sure that overproduced electricity is consumed.

To make sure that the cables are not overloaded we include capacity constraints for each phase. This means that for each group of houses (connected to a phase) the amount of power transported is not allowed to be higher than some capacity. For the transformer we make sure that the amount imported to and exported from the district is not higher than the transformer's capacity.

Future electronic equipment such as electric vehicles are expected to be widely used in the near future. These electronic devices will create the need for larger grid capacity and have quite a volatile electricity demand. This can cause problems on the electricity grid. In this chapter we only consider electric vehicles and heat pumps. We model several amounts of electric vehicles and heat pumps to find out whether they can be complemented by some combination of DGs. If we want to include the use of electric vehicles in the model we need to make some assumptions. Firstly, we assume that electric vehicles have a capacity of 15 kW and need to be charged for five hours. Secondly, we assume that electric vehicles are only charged at home. And thirdly, we assume that on average the electric vehicles are charged at 19:00. In this chapter we only consider heat pumps that rely on electricity. These heat pumps have higher electricity demand and will thus have a bigger impact on the model. The base demand profile of heat pumps

4.2 Optimal mix of distributed generators

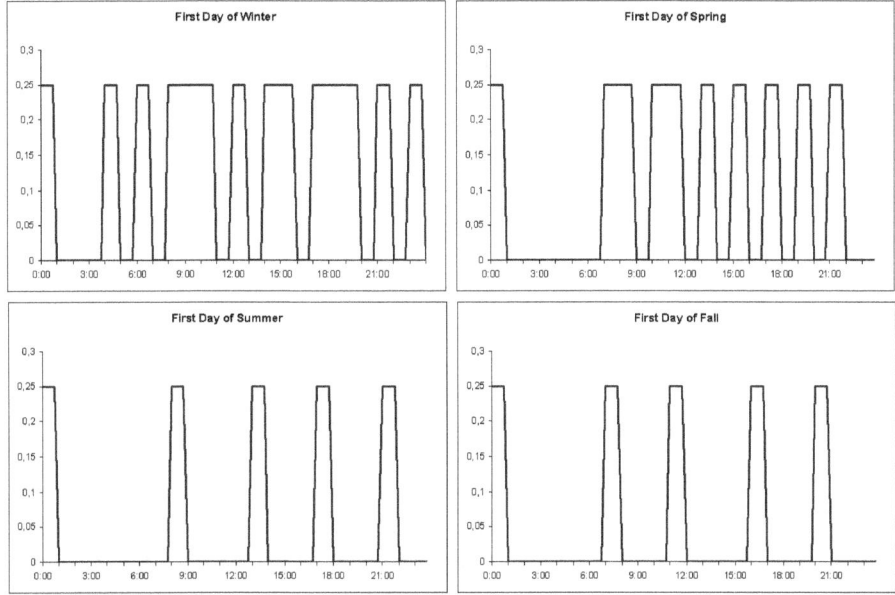

Figure 4.3: Heat pump daily demand profiles for each season.

is plotted in Figure 4.3. From these graphs we see that the electricity demand of heat pumps is very variable, especially in winter and spring. Because of this variability it will be difficult to perform Monte Carlo simulations on the demand data. Because heat pumps constantly turn on and off in winter and spring, we obtain profiles of heat pumps that continuously consume electricity in these seasons.

4.2.1.2 Power flow and DG variables

To match supply and demand there are a few different possibilities as to who supplies electricity to whom. We have the following entities: 'power plants', 'district', 'other districts', 'storage system' and 'house'. The power flows between these entities are depicted in Figure 4.4. The entity 'power plants' can only supply, 'district' can supply and consume power, 'other districts' can only consume, 'storage system' can retain and release, and 'house' can supply and consume.

All variables representing power flows are denoted by $P_{..}$ where the indices correspond to the following entities: 'p' stands for power plants, 'o' stands for other districts, 'd' stands for district, 's' stands for storage, and 'h' stands for house. This means that the amount of electricity imported 'from power plants to the district' is denoted as P_{pd}. Similarly, for power flows 'from the district to other districts' are denoted as P_{do}, power flows 'from storage to district' and 'district to storage' are denoted as P_{sd} and P_{ds}, respectively, and for 'district to house i' and 'house i to district', $P_{dh,i}$ and $P_{hd,i}$, respectively, where $i = 1, \ldots, n$ and n is the number of houses in the district.

The demand and production of electricity are measured in time periods. Conse-

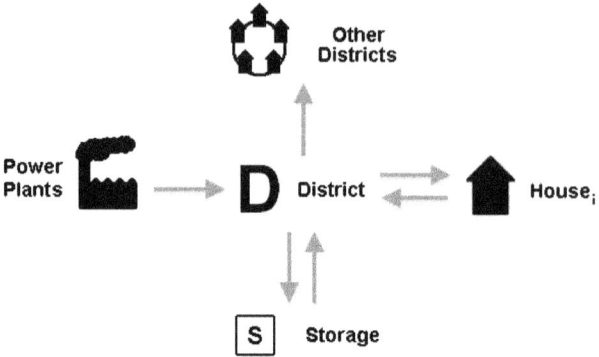

Figure 4.4: Power flows in mathematical model.

quently, the power flows in the district are calculated for each time period. To include this in the model, we add an extra index $t \in \{1, \ldots, T\}$ which stands for the time period $[t-1, t)$. So all variables above representing power flows have this extra index. We have the following variables P_{pd}^t, P_{do}^t, P_{sd}^t, P_{ds}^t, $P_{dh,i}^t$, and $P_{hd,i}^t$ denoting power flows at time t.

The decision variable indicates whether DG type j is present in house i or not. This means that we have binary decision variables for each house and each DG type. Suppose that we have n houses and m DG types, then for $i = 1, \ldots n$ and $j = 1, \ldots, m$ we have

$$DG_{i,j} = \begin{cases} 1 & \text{if house } i \text{ has DG type } j, \\ 0 & \text{otherwise.} \end{cases}$$

4.2.1.3 Energy loss

The objective is to minimize the overall energy loss. To get a complete view of the grid, we include:

- the loss of transporting electricity from power plants all the way to the houses in the district.

- the loss of exporting electricity from the houses in the district to houses in other districts.

- the loss of transporting within the district, i.e. (i) from one house to another and (ii) between the houses and the storage system.

- the loss of using the storage system.

In this subsection we go into more detail in these categories and how to calculate them.

4.2 Optimal mix of distributed generators   137

4.2.1.3.1 Introduction to loss

The loss of transporting electricity consists of losses in cables and/or losses in transformers. Cable losses have to do with the heat generated in cables. The higher the resistance in cables the higher the losses. As different materials have different resistances, cable losses depend on the material used as conductor. In transformers energy is dissipated in the windings, core, and surrounding structures. Transformer losses are divided into losses in the windings, termed copper loss, and those in the magnetic circuit, termed iron loss.

When using the storage system energy loss is created due to self-discharge and inefficiency. The latter one is the loss of transforming electricity into e.g. mechanical or chemical energy (depending on the storage technology), and then back again into electric energy.

There is limited data on the size of network losses. The total network losses are estimated by [43] at 7 to 8 % of the power consumed for the entire chain, i.e. from power plants to consumers. The U.S. Energy Information Administration (EIA, http://www.eia.gov) reports an estimation of 7% on their website. To find out whether incorporating DGs in the district decreases energy loss, we need to make some assumptions on the power losses in High-, Middle-, and Low-Voltage grids.

As mentioned above, the loss of transporting electricity from power plants to houses is between 7 and 8%. The loss of exporting electricity to outside the district is calculated by adding the losses in the LV-grid, losses in the MV/LV-transformer, losses in the MV-distribution grid, losses in the other MV/LV-transformer, and losses in the other LV-grid, which is estimated at around 6.9%. The average loss percentage for transporting within the district is estimated at around 1.1%. These percentages are averages and exclude the effect of temperature changes and different conductor materials.

	Lead-acid	NaS	Zinc-bromide	Vanadium-redox
Power (kW)	up to 20,000	100 - 30,000	10 - 250	5 - 500
Energy (kWh)	1 - 40,000	1 - 50,000	50 - 500	50 - 2000
Lifetime (cycles)	200 - 1200	1000 - 4000	1000	5000 - 12,000
Cycle Efficiency	75 - 80%	85 - 90%	65 - 75%	80 - 87%
Response Time	30 ms	30 ms	30 ms	30 ms
Storage Time	6-8 hrs	±1 hr	±2 hrs	-
Self Discharge	2 - 5%/month	nil[2]	nil	nil
(Dis)charge time	0.5 - 5 hrs	8 hrs	0.5 - 3 hrs	1 hour
Density (Wh/kg)	20 - 50	100 - 200	70 - 90	20 - 40
(W/kg)	75 - 300	150 - 250	60 - 140	180
Cost (€/kWh)	200 - 900	225 - 400	1500	100 - 500
(€/kW)	45 - 450	300	1500	4000 - 10,000
Environment	lead disposal, H2	chemical handling	none	benign
Development	commercial	in development	in test	in development

Table 4.1: Characteristics of different storage systems.

For the calculation of energy loss from using the storage system, the efficiencies and self-discharge rates for different storage technology types are reported in Table 4.1, based on data from Schoening [133], Hadjipaschalis et al. [59] and Lysen [100]. It is expected that in the near future these technologies will be more developed and have higher efficiencies. We model a (not yet existing) storage system with a low self-discharge rate and a high efficiency. This futuristic storage system will be referred as StorageX, and we choose a self-discharge rate of 2% per month and an efficiency of 95%. We model scenarios of different self-discharge rates and efficiencies to find out how efficient a storage system should be so that it will be used in the district and how different storage efficiencies affect the optimal mix of DGs.

4.2.1.3.2 Calculating transportation loss
Electric power depends on voltage and current as is shown in the following equation

$$P = I \cdot V,$$

where P represents the electric power (in Watt), I represents the electric current (in Ampere) and V represents the potential difference (in Volt). This equation implies that, keeping voltage the same, a higher amount of power requires a higher current flow. Joule's law states that

$$Q = I^2 \cdot R \cdot t,$$

where Q is heat generated (in joule) for a time t. This equation implies that, if resistance is kept the same, higher current levels lead to higher generated heat. And as heat is power loss, higher current causes more power loss. In a nutshell, more power leads to higher current (keeping voltage the same), and as a result more heat is generated and thus leads to higher energy loss.

These equations show that energy loss has a quadratic relation to load: the higher the load the higher the relative loss. So we cannot simply multiply the average loss percentages (7.5%, 6.9%, and 1.1%) with the power flows, because then the quadratic relation between loss and load will not be present in the model. Furthermore, these percentages are for the electricity grid without distributed generation. We deal with these issues as follows:

(1) To include the quadratic relation between loss and load, one should quadratically normalize the power loss percentage for each day whilst making sure that it still equals the average percentage. The reason to normalize it for each day is because the daily consumer profile does not change much. Define v_t as the load for time period t and $\ell \in [0,1]$ as the average loss fraction. Then the power loss w_t at time t is calculated as

$$w_t = v_t^2 \cdot \ell \cdot \frac{\sum_{t \in day} v_t}{\sum_{t \in day} v_t^2}.$$

(2) The obtained average loss percentages are for the current grid. But because we model a district with DGs, these average percentages of loss in the grid can turn out to be higher or lower. This follows from the quadratic relation between loss

4.2 Optimal mix of distributed generators

and load. Since we do not know the average loss percentages in a grid with DGs we will first normalize the load in the grid with DGs (x_t) by the load in the current grid (v_t) for each time period t. Then we multiply it by the loss as calculated in the current grid. So the power loss in the grid with DGs y_t at time t is calculated as:

$$y_t = \left(\frac{x_t}{v_t}\right)^2 \cdot w_t = \left(\frac{x_t}{v_t}\right)^2 \cdot \left(v_t^2 \cdot \ell \cdot \frac{\sum_{t \in day} v_t}{\sum_{t \in day} v_t^2}\right)$$

$$= x_t^2 \cdot \ell \cdot \frac{\sum_{t \in day} v_t}{\sum_{t \in day} v_t^2}.$$

4.2.1.3.3 Incorporating transportation loss

In the method discussed above the transportation loss is calculated for each time period, where the load is multiplied by $\ell \cdot \frac{\sum_{t \in day} v_t}{\sum_{t \in day} v_t^2}$. This term depends on the total sum of load in the corresponding day. As a result we obtain a different loss coefficient for each day, leading to a large objective function containing many loss coefficients. While having only one loss coefficient for importing from power plants, one for exporting to other districts and one for transporting within the district leads to a much simpler and smaller objective function.

Since loss has a quadratic relation to load, we will use quadratic regression to obtain these loss coefficients. The methodology is explained as follows. Using a simulation model, which calculates transportation loss using the method explained earlier, we calculate the loads and losses in four scenarios: the current scenario (the grid without DGs), the quarter mix scenario (a quarter of the houses have all DGs), the half mix scenario (half of the houses have all DGs) and the maxed out scenario (all houses have all DGs). In each scenario we obtain loads and losses of (i) importing electricity from power plants, (ii) exporting to other districts, and (iii) transporting within the district. To obtain the loss coefficient for each of these three cases we apply quadratic regression on the corresponding loads and losses.

We start by estimating the loss coefficient of importing electricity from power pants. The resulted load data of all scenarios and all time periods are put into one vector $\mathbf{x}$ to represent the different loads of import over a whole range of scenarios, from no DGs to a lot of DGs in the district. The same is done for the resulted loss data which are put into vector $\mathbf{y}$ representing the loss data corresponding to the loads of import in $\mathbf{x}$. Then we have the following quadratic regression model:

$$\mathbf{y} = \ell_p \mathbf{x}^2 + \epsilon, \tag{4.1}$$

where ℓ_p is the loss coefficient of importing from power plants and ϵ is the residual vector. The estimated ℓ_p is used in the optimization model to calculate losses from importing electricity into the district.

The same methodology is carried out to estimate the loss coefficient of exporting electricity to other districts (ℓ_o) and the loss coefficient of transporting electricity within the district (ℓ_d). We have obtained the following estimated quadratic coefficients of loss: for import ($\hat{\ell}_p$) it is 0.00233, for export ($\hat{\ell}_o$) it is 0.00229 and for within district ($\hat{\ell}_d$) it is 0.00035. For comparison, both the 'real' and estimated losses are plotted in Figure 4.5.

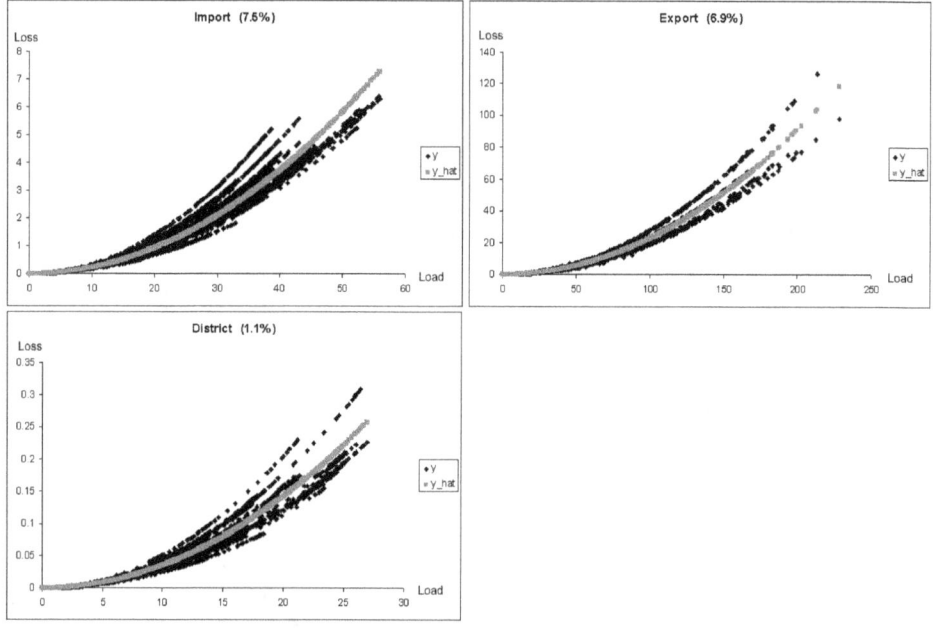

Figure 4.5: Quadratic regression of loss and load in kWh.

The problem with this approach is that the estimated losses of transporting electricity may not be proportional to the losses of using the storage system, i.e. if we underestimate the transportation losses then it becomes less efficient to use the storage system, while if we overestimate the transportation losses then it becomes more efficient to use the storage system.

The way losses are estimated is a bit ad hoc, but there is very little data available. One can improve it by including technical electrical aspects which is beyond the scope of this thesis.

When electricity is transported from one place to another, there is energy loss. So the amount of electricity that is supplied is not the same as the amount received. This means that each time electricity is transported, an additional amount of electricity needs to be supplied to compensate the loss. But transporting this additional amount of electricity will also create energy loss. So again an additional amount of electricity needs to be imported, which again creates loss. To keep it simple we exclude these losses from the balance constraints in the model and only incur them in the objective function.

4.2.1.4 Constraints

For each house the demand has to be satisfied. The supply of electricity can come from power plants, its own generated electricity, overproduction of other houses or from the storage system. In addition overproduced electricity must be stored or exported to other districts. There are also some restrictions that we need to consider. These are the capacity constraints of cables, transformer and storage system.

4.2.1.4.1 Supply and Demand-balancing constraints

In the system, all the demand of all houses in the district has to be satisfied. Each house can use electricity from their own production. If there is not enough generated electricity then a house can import from the main grid. Or if there is overproduction they can deliver it back to the main grid. Then for each house i at time t we have the following constraints

$$\sum_{j=1}^{m} s_{i,j,t} \cdot DG_{i,j} + P^t_{dh,i} - P^t_{hd,i} = d_{i,t}, \qquad (4.2)$$

where d^t_i is the demand of house i at time t and $s_{i,j,t}$ stands for the generated electricity of house i from DG type j at time t. These constraints make sure that the demand of each house is satisfied by its own DGs and/or the main grid.

Now the supply/demand-balancing constraints can be defined for the whole district. The main grid can receive electricity, aside from power plants, also from the storage system and houses. And the other way around, the main grid provides electricity to houses, storage system and other districts. To satisfy the demand of the whole district at time t, we have the following constraints:

$$\sum_{i=1}^{n}\sum_{j=1}^{m} s_{i,j,t} \cdot DG_{i,j} + P^t_{pd} - P^t_{do} + P^t_{sd} - P^t_{ds} = \sum_{i=1}^{n} d_{i,t}. \qquad (4.3)$$

So the demand of the whole district ($\sum_i d_{i,t}$) must be equal to the district's production ($\sum_{i,j} s_{i,j,t} DG_{i,j}$), plus import ($P^t_{pd}$) and/or from storage ($P^t_{sd}$) if there is underproduction, and minus export (P^t_{do}) and/or to storage (P^t_{ds}) if there is overproduction. Whether the storage system is used depends on how efficient the storage system is compared to importing and exporting. Constraints (4.2) and (4.3) imply

$$\sum_i (P^t_{dh,i} - P^t_{hd,i}) = (P^t_{pd} - P^t_{do}) + (P^t_{sd} - P^t_{ds}),$$

that is, the net flow of electricity from houses should be equal to the net flow of electricity from import/export plus the net flow of electricity from the storage system.

4.2.1.4.2 Capacity constraints

For the transformer the following capacity constraints hold for each time period t

$$P^t_{pd} + P^t_{do} \leq cap_{tra}. \qquad (4.4)$$

So the amount imported or exported is not allowed to be higher than the capacity cap_{tra}. For cables the houses are split into groups, as discussed earlier. The following cable constraints hold for each group of houses g and time period t

$$\sum_{i \in A_g}(P^t_{dh,i} + P^t_{hd,i}) \leq cap_{cab}, \tag{4.5}$$

where A_g is the set of indices for each group of houses $g \in \{1,\ldots,G\}$ connected to one phase. So the amount of electricity transported to or from a group of houses is not allowed to be higher than the maximum capacity cap_{cab}.

4.2.1.4.3 Storage constraints

Storage systems have restrictions on energy and power capacity. Power is expressed in kilowatts (kW) and energy is expressed in kilowatt-hours (kWh). The capacity at which the storage system can store or release energy is the power capacity and the maximum amount of energy that can be stored is given by the energy capacity. Define cap_{ch} as the power capacity, then the charge/discharge constraint for each time period t is

$$P^t_{ds} + P^t_{sd} \leq cap_{ch}. \tag{4.6}$$

Define $\ell_s \in [0,1]$ as the self-discharge rate, $E \in [0,1]$ as the efficiency of the storage system, B_0 as the storage level at the starting period ($t = 0$) and cap_{stor} as the energy capacity, then the storage capacity constraints for each time period $t \in T$ are

$$\sum_{k=1}^{t}\left[\left(E \cdot P^k_{ds} - P^k_{sd}\right)(1-\ell_s)^{t-k}\right] + B_0(1-\ell_s)^{t-1} \geq 0, \tag{4.7}$$

and

$$\sum_{k=1}^{t}\left[\left(E \cdot P^k_{ds} - P^k_{sd}\right)(1-\ell_s)^{t-k}\right] + B_0(1-\ell_s)^{t-1} \leq cap_{stor}. \tag{4.8}$$

The storage capacity constraints basically state that the storage level is not allowed to be negative or higher than some capacity cap_{stor}. The storage level at time t is calculated by summing the amount of energy that is retained and release from the starting period till time t, discounted for the loss from self-discharge.

4.2.1.4.4 Binary and non-negativity constraints

As discussed in Section 4.2.1.2, the DG-variables are binary. So for each house i and DG type j, we have

$$DG_{i,j} \in \{0,1\}. \tag{4.9}$$

In addition, non-negativity constraints are included for the power flow variables. For each time period t

$$P^t_{pd}, P^t_{do}, P^t_{sd}, P^t_{ds} \geq 0, \tag{4.10}$$

and for each house i and time t

$$P^t_{dh,i}, P^t_{hd,i} \geq 0. \tag{4.11}$$

4.2 Optimal mix of distributed generators

4.2.1.5 Objective function

The objective is to minimize loss of importing, exporting, transporting within district, and using the storage system. Firstly, the quadratic terms of the objective function are discussed, i.e. losses from transporting electricity, and secondly, the linear terms, i.e. losses from using the storage system.

Define $\hat{\ell}_p$, $\hat{\ell}_o$ and $\hat{\ell}_d$ as the estimated coefficients from the quadratic regression model (4.1) for 7.5%, 6.9% and 1.1% average loss, respectively. Then for the quadratic part applies

$$\Gamma = \sum_{t=1}^{T} \left(\hat{\ell}_p {P_{pd}^t}^2 + \hat{\ell}_o {P_{do}^t}^2 + \hat{\ell}_d \left(\sum_{i=1}^{n} P_{hd,i}^t - P_{do}^t + P_{sd}^t \right)^2 \right). \qquad (4.12)$$

The first quadratic term is the loss of importing, the second quadratic term is the loss of exporting and the last quadratic term is the loss of transporting within the district. The first two terms are pretty straightforward. The last term is explained as follows. A power flow from house i to the district, $P_{hd,i}^t$, consists of transporting from house i to (i) other houses in the district, (ii) the storage system and (iii) other districts as export. Since the loss of exporting to outside the district is already incurred in the objective function, the amount exported should be subtracted. Then still the power flows from the storage system to the district has to be added.[3]

The linear part that describes the loss of using the storage system is calculated as

$$\Delta = \sum_{t=1}^{T-1} \ell_s \left(\sum_{k=1}^{t} \left(E \cdot P_{ds}^k - P_{sd}^k\right)(1-\ell_s)^{t-k} + B_0(1-\ell_s)^{t-1} \right) + \sum_{t=1}^{T}(1-E)P_{ds}^t, \qquad (4.13)$$

where $\ell_s \in [0,1]$ is the self-discharge rate, $E \in [0,1]$ the efficiency of the storage system and B_0 the storage level at the starting period ($t=0$). The first term represents the loss from self-discharge, where the term between the large brackets is the storage level. The second term is the loss from converting electricity from one form to another, i.e. the inefficiency of the storage system.

4.2.1.6 Overview of the mathematical model

The optimization model is a Mixed Integer Quadratic Programming (MIQP) problem, due to some binary variables and quadratic objective function. Here a complete overview is given of the mathematical model with its objective function and constraints.

$$\min \Gamma + \Delta + \sum_{t=1}^{T}(1-E)P_{ds}^t,$$

---

[3]This can also be done the other way around: subtract the flow of imports from the flows of district to houses and then add the flow of district to storage. In mathematical terms: $\sum_{i=1}^{n} P_{dh,i}^t - P_{pd}^t + P_{ds}^t$. This is equal to the previous terms $\sum_{i=1}^{n} P_{hd,i}^t - P_{do}^t + P_{sd}^t$. These equations can be deducted from the two supply/demand-balancing constraints (4.2) and (4.3).

subject to the following constraints:

$$\sum_{j=1}^{m} s_{i,j,t} \cdot DG_{i,j} + P_{dh,i}^t - P_{hd,i}^t = d_{i,t} \quad (i=1,\ldots,n \quad t=1,\ldots,T),$$

$$\sum_{i=1}^{n}\sum_{j=1}^{m} s_{i,j,t} \cdot DG_{i,j} + P_{pd}^t - P_{do}^t + P_{sd}^t - P_{ds}^t = \sum_{i=1}^{n} d_{i,t} \quad (t=1,\ldots,T),$$

$$P_{pd}^t + P_{do}^t \leq cap_{tra} \quad (t=1,\ldots,T),$$

$$\sum_{i \in A_g} (P_{dh,i}^t + P_{hd,i}^t) \leq cap_{cab} \quad (t=1,\ldots,T \quad g=1,\ldots,G),$$

$$P_{ds}^t + P_{sd}^t \leq cap_{ch} \quad (t=1,\ldots,T),$$

$$\sum_{k=1}^{t} \left(E \cdot P_{ds}^k - P_{sd}^k\right)(1-\ell_s)^{t-k} + B_0(1-\ell_s)^{t-1} \geq 0 \quad (t=1,\ldots,T),$$

$$\sum_{k=1}^{t} \left(E \cdot P_{ds}^k - P_{sd}^k\right)(1-\ell_s)^{t-k} + B_0(1-\ell_s)^{t-1} \leq cap_{stor} \quad (t=1,\ldots,T),$$

$$DG_{i,j} \in \{0,1\} \quad (i=1,\ldots,n \quad j=1,\ldots,m),$$

$$P_{pd}^t, P_{do}^t, P_{sd}^t, P_{ds}^t \geq 0 \quad (t=1,\ldots,T),$$

$$P_{dh,i}^t, P_{hd,i}^t \geq 0 \quad (i=1,\ldots,n \quad t=1,\ldots,T).$$

4.2.2 Description of data

The DGs that are considered are micro-CHP systems, PV solar panels and micro wind turbines. The amount of electricity that each of these DGs will generate is unknown, but one can make some assumptions and use the characteristics of these DGs to make average profiles. Micro-CHP systems depend on heat consumption, PV solar panels on sunlight and micro wind turbines on wind speed. The production of electricity by micro-CHP systems and PV solar panels is variable but still quite predictable. On the other hand, the production of micro wind turbines is very unpredictable. This is because of the intermittent nature of wind in urban areas and the different effects that obstacles, such as buildings and trees, have on wind speed and direction. The data for these three DGs are discussed in the following subsections. Monte Carlo simulations are performed on the DGs' production profiles.

The demand and production data are reported for every 15 minutes and are split into four seasons: winter, spring, summer and fall. We are modelling only one week per season so that the data set does not become too large while still taking the seasonal effect into account. So each season is represented by one week and starts on Wednesday. Wednesday is chosen as the first day of the week because it represents an average day without the influence of weekends.

Since we are only modelling one week for each season we need to make some assumptions for the storage system as the amount of electricity stored at the end of each week will probably not be the same as the amount of electricity stored at the beginning of the week in the following season. But when optimizing the use of the storage system it is important that we allow the storage to store and release energy freely without interruptions between seasons. So we assume that the storage system at the end of each week is the same as the storage level in the beginning of the week in the following season.

There is only limited data available to us. For demand of a household and production by DGs there is no real data available. But we can make some general assumptions on household demands and DGs' performances. This way we can obtain average profiles from which we can make some general conclusions. We describe the assumptions and data that we used in the example of this chapter. The model can be used with real data when available.

The main focus is on a group of houses in a district. But as there is no one district that represents all districts and one house that represents all houses, we need to make some assumptions on the district composition and the different types of houses. We have no data available on the demands of households. But as most houses have a fairly predictable demand profile we can use this as a tool for modelling the houses' demands.

Our definition of a district is a group of houses connected to an MV/LV transformer (see Figure 4.2). We assume that there are 250 houses in the district. Five types of houses will be considered: detached, semi-detached, terraced, apartment and maisonette. The average demands and shares of houses are reported in Table 4.2, where the average demand is the average year demand typical for the house type and the share of houses is the overall share of each house type in the Netherlands. Using these data we can construct a district that represents the average composition of houses in the Netherlands.

We start with one average demand profile, which is the base demand, which is given

	Demand	Share
Detached	5000	14.8
Semi-Detached	4000	12.4
Terraced	3500	42.6
Apartment	3000	25.5
Maisonette	3500	4.7

Table 4.2: Average demand in kWh and share for each house type.

as a percentage of the total average year demand for every fifteen minutes. This base demand profile is used to model the demand for all houses. To create different variants of demand for each house we perform Monte Carlo simulations on the base demand. The Monte Carlo simulations are performed as follows. Electricity demand fluctuates a lot during the year due to seasonal changes. As explained earlier, we split a year into four seasons and for every season we model one week.

4.2.3 Solution method

The Mixed Integer Quadratic Programming (MIQP) problem of the previous section can be reformulated into the standard matrix form. Let H be a symmetric matrix describing the coefficients of the quadratic terms in the objective function, f the parameter vector describing the coefficients of the linear terms in the objective function, A and b the matrix and vector corresponding to the inequality constraints, and Aeq and beq the matrix and vector for the equality constraints. The variable x is a $(4T + 2nT + nm)$-vector, where T, n and m are the number of time periods, houses and DG types, respectively. Let $\mathcal{C}$ be the set of indices of the continuous variables of x and $\mathcal{B}$ the set of indices of the binary variables. Then the quadratic optimization model can be formulated as

$$
\begin{aligned}
\text{minimize} \quad & fx + x'Hx \\
\text{subject to} \quad & A\,x \leq b \\
& Aeq\,x = beq \\
& x_c \geq 0, \quad c \in \mathcal{C} \\
& x_b \in \{0,1\}, \quad b \in \mathcal{B}.
\end{aligned}
\tag{4.14}
$$

MIQP problems are in general hard to solve. We used two methods to simplify the problem: we reduced the problem size and relaxed the binary variables so we now have a QP problem. This problem is well solvable by AIMMS and the CONOPT solver.

4.2.3.1 Reduce problem size

If we try to solve the current model we would have $4T + 2nT + nm = 1{,}355{,}252$ variables and $7T + 3nT + GT + nm = 2{,}075{,}636$ constraints. With the given computer power we are not able to solve such a large model. So we need ways to reduce the problem size. We can:

- Transform 15-minute data into hourly data.

- Model one day instead of one week for each season. We only model Wednesdays as these are average days without the influence of weekends.

- Decrease the number of houses. We can, for example, model 25, 50 and 100 houses. This way we see the effect of larger districts and can ultimately make general conclusions for large districts without having to actually solve large models.

The drawback is that because we model fewer time periods and fewer houses, and aggregate the data into hours (instead of 15 minutes) our solutions will be less accurate. But we implement all of them so that the problem at least becomes solvable. These reduction measures lead to a model with 48,884 variables and 74,612 constraints for a district of 250 houses.

4.2.3.2 Relax binary variables

Recall that the feasible set is not convex due to binary variables. These binary constraints can be relaxed such that the DG variables of micro-CHPs and PV solar panels are in the interval [0,1]. DG-variables having values lower than 1 are interpreted as DGs with lower production capacities. In the case of PV solar panels, this is interpreted as having fewer solar panels on the roof. PV solar panels can be between $1m^2$ to $8m^2$. Recall that our data is for an $8\ m^2$ PV. One drawback of relaxing the binary constraints is that it will probably be more efficient if all houses have small amounts of DG production. Since investing in DG systems with too low production profiles can become unprofitable, we may end up with a solution that cannot be implemented.

After relaxing the binary variables a QP model is obtained. QP problems are less complex and thus easier to solve. Since the square matrix H in the objective of Model (4.14) is positive semidefinite[4], the objective function is convex and there is a global minimizer if there is a feasible solution. With QP problems one can use the Karush-Kuhn-Tucker conditions to find optimal solutions.[5] But since the parameter matrix H in the objective is not positive definite, the Karush-Kuhn-Tucker conditions are necessary but not sufficient. Hence, these conditions are not sufficient to find optimal solutions.

A positive semidefinite square matrix H allows the existence of multiple optimal solutions. Having multiple optimal solutions in power flow variables is not a problem as we are trying to find an optimal mix of DGs. In addition, having multiple optimal solutions in DG variables is not necessarily a problem since multiple optimal solutions gives the decision maker opportunity to choose an optimal combination based on additional criteria.

Notice that we are not trying to find an optimal allocation of DGs for the houses, but more a share of DGs for the district as a whole. This means that if house A has a micro-CHP in an optimal solution, while there is another optimal solution where house B has a micro-CHP instead of house A, then for our purpose this does not matter as the share of DGs in the optimal solution stays the same.

---

[4] A real symmetric matrix M is positive semidefinite if $z^T M z \geq 0$, $\forall z$.

[5] For more information on the Karush-Kuhn-Tucker conditions one can refer to the literature, e.g. Peressini et al. [116]

4.2.4 Results

In this section the main results obtained from the optimization model are discussed. The model was solved with and without the possibility to export, with and without a storage system, and for different amounts of electric vehicles and heat pumps, and gives us a huge amount of results. A full overview of these results and analyses can be found in [33].

To get some feeling for the calculation of loss and energy flow of our model we tested it with four scenarios: the No DGs scenario, the quarter mix scenario (a quarter of the houses have both micro-CHPs and PV solar panels), the half mix scenario (half of the houses have both micro-CHPs and PV solar panels) and the maxed out scenario (all houses have both micro-CHPs and PV solar panels). In each scenario we calculate the production and the amount consumed for each house from the main grid. Then we calculated the total import, export and the amount transported within the district. From this we calculate losses for every time period. In Figure 4.6 we plot the total energy loss for each scenario. This is the same graph as we saw earlier by Scheepers and Wals [132] but now using the results from our simulation model. With the optimization model we want to find the lowest point in the graph by vitiating both percentages of the DGs.

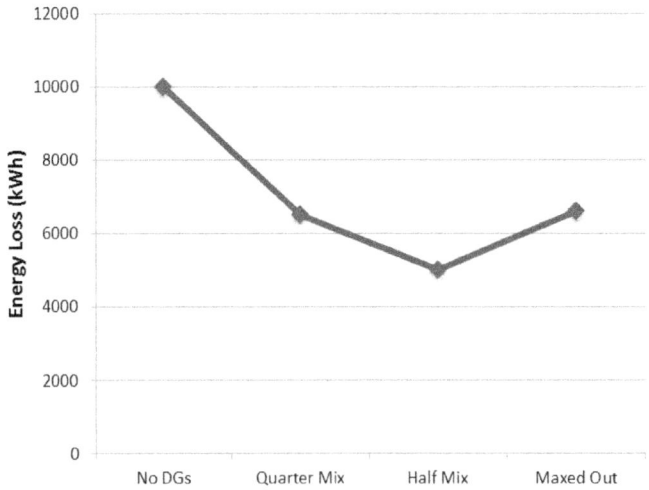

Figure 4.6: Loss in scenarios.

From all the results we generated, we need to extract recommendations on the mix of DGs that should be implemented in a district. To this end it is useful to distinguish four periods: the first period is now the current grid without DGs; the second one is the transition period, where DGs generate electricity; the third is also in the future, where DGs generate and more usable storage systems exist and the last one is even more into the future, where there is an efficient storage system and extra demand from electric vehicles and heat pumps. In all scenarios we assume export to other districts is not

4.2 Optimal mix of distributed generators

possible.

The characteristics of these periods with the optimal mix of DGs are reported in Table 4.3. Notice that over time more DGs are needed in the district until it leads to all houses having both types of DGs. This follows from the high demand of electricity when using electric vehicles and heat pumps.

	Current period	Transition period 1	Transition period 2	Future period
DGs	-	+	+	+
Export	-	-	-	-
Storage	-	-	+	+
HPs & EVs	-	-	-	+
Micro-CHP	low	23%	94%	100%
PV	low	54%	100%	100%

Table 4.3: Characteristics of the periods with optimal mix.

The storage that is assumed here is a future high efficient storage system. With the current known storage systems the percentage of micro-CHP varies from 38% (in low efficient storage systems like flywheel) to 80% (in highly efficient storage systems like NaS). The percentage of PV will vary between 75% and 100%.

The introduction of (the optimal mix of) DGs and the storage possibility reduces the average loss per household. Following from our model energy losses are reduced from 13.4 kWh per household in the current situation to 7.8 kWh by introducing the optimal mix of DGs and to 5.7 kWh by the addition of an efficient storage system in combination with the optimal mix of DGs in that situation. This is depicted in Figure 4.7.

Because we use average data and an average composition of districts, one should not simply implement our optimal mix of DGs and expect that it is the best one in all situations. All districts have different compositions and different demand patterns. Our results are for the average case. However, we have shown that these solutions give a large decrease in total loss and can thus be used as a guideline for incorporating DGs in a district.

All optimal mixes in the scenarios without heat pumps and electric vehicles are feasible solutions. The cable capacity constraints are not violated. In the case of heat pumps and electric vehicles[6] in most solutions the cable and transformer capacity constraints were violated. To get a better understanding of the results we consider the following four cases: (i) one heat pump in the district, (ii) one electric vehicle in the district, (iii) two heat pumps in the district on the same phase, and (iv) one heat pump in each phase (there are 15 phases so 15 heat pumps are included in the district). The results obtained are the following:

(i) One heat pump: total loss is 692 kWh with 94% of micro-CHP systems and 100% of PV solar panels,

(ii) One electric vehicle: infeasible,

---
[6]We do not mean here the current vehicles with a very limited action radius on the battery

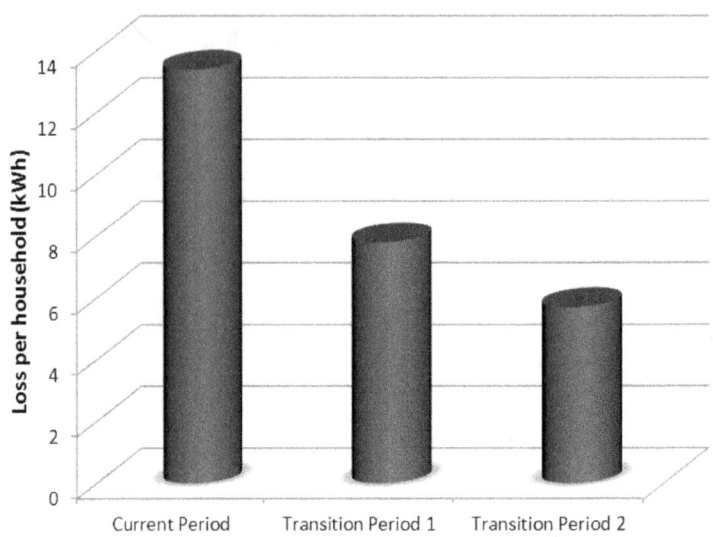

Figure 4.7: Saving due to efficient mix and storage system.

(iii) Two heat pumps on the same phase: infeasible,

(iv) One heat pump in each phase: total loss is 758 kWh with 98% of micro-CHP systems and 100% of PV solar panels.

So, in the district with current cable and transformer capacities it is not possible for houses to have electric vehicles. Also, only one house on each phase is allowed to have a heat pump. This is very unfortunate, but these future electronic devices have too high demands and will overload transformers and cables. Hence, if one wants to include electric vehicles and heat pumps in a district, the cables and transformers must be reinforced to be able to handle such a large increase in demand.

4.2.5 Implementation

When developing new districts one can simply use our solutions to deploy DGs in the district. Because these houses do not have an owner yet, DGs can be simply distributed using the optimal mix. In addition, the cost from installing these DGs can be incorporated into the price of houses. Implementing our solutions in already existing districts will be a bit more tricky. Most people do not want to take the risk to invest in new technologies unless there is some financial assistance from the government, like subsidies. But the budget for these subsidies is quickly exhausted, leaving some empty handed. By stimulating collaboration within districts one can reduce the barrier to invest in DGs. Home owners as a group may feel less intimidated by big investments, especially if governments also provide some financial assistance. See Hauge et al. [61] for an example.

A district would need the collaboration of all home owners where all must agree to comply with the optimal share of DGs allowed in the district. As many projects to increase involvements in the community have failed, we realize that to make such a collaboration between many home owners will be difficult. But one can implement optimal solutions on a smaller scale between home owners that already (or want to) work together. In the Netherlands, there are cases where groups of people collaborate to invest in DGs, such as solar panels and wind turbines, even without the support from the government. However, this leads to another problem: using our optimal solution, some home owners may end up without a micro-CHP system. This makes the implementation more complicated because how does a group of home owners decide which household will not have a DG.

Another way to implement our solutions is to let electric utilities or other companies rent rooftops or spaces to place DGs.[7] So instead of letting home owners buy DGs, companies can invest in these technologies. This way they can implement the most efficient solution and choose where they want to put a DG system. Whether higher efficiency implies higher profit is a good question for further research. One way is to replace energy loss with costs in the objective function and include other cost aspects, such as the prices of DGs and the price of electricity.

There are some examples where companies start to rent rooftops to place their solar panels. In Europe, for example, there is a company called GreenPulse that rents rooftops of large buildings to generate electricity from solar panels (http://www.greenpulse.eu). In the US there are some states where companies, such as Pure Energies, rent residential rooftops to place their solar panels (http://www.pure-energies.com).

4.3 Optimal placing of wind turbines

When looking at the optimal place to locate a wind turbine, trade-offs have to be made between local placement and spreading: transmission loss tends to local placements and the correlation between the stochastic productions of wind turbines tends to spreading. In this chapter steps are described to determine the locations of new wind turbines that minimize energy loss on the High Voltage power grid. A vindication of the used power grid model is provided, the simulation procedure for stochastic wind power is described and the required mathematical optimization models are described as well as implemented. Results are shown and their relation to real life problems is discussed.

4.3.1 Power grid modelling

In this section an introduction into power grids is given. Then the power flow equations are formulated, energy loss is described, and by several steps a simplified model for power grid behaviour is derived.

4.3.1.1 Power grids

A *power grid* is a set of electrical components called *buses*, which are connected with each other by transmission lines, called *branches*. Some of the buses represent electricity

---

[7]In the case of solar panels companies can simply rent rooftops and pay a fee to the home owners. But for micro-CHP systems this is more complex as production depends on heat consumption.

consumers, which have a demand for electrical power. These buses are called *load buses*. Some other buses represent power plants, or other sources of electrical power. These buses are called *generation buses*.

In order to keep the voltages through the entire grid at their demanded level, the system is operated in such a way that supply equals demand at every point in time. As demand varies in time depending on the wishes of the consumers, power generation has to be adjusted accordingly.

Due to resistance of transmission lines, electrical energy is lost during transportation from generation to load buses. These losses are reduced by applying higher voltages, but electricity has to be delivered to consumers at only $230V$. As a result, most power grids are designed in the following way (see also Figure 4.2):

- **High Voltage grid**, or *transmission grid*. It is kept at voltages of $110kV$ to $450kV$, it contains all large power plants, and some connections abroad. It was implemented as a number of interconnected double rings.

- **Medium Voltage grid**, being the link between transmission and distribution. It is kept at voltages between $1kV$ and $110kV$, it is connected to the HV grid by so called *transmission stations*, and it contains large consumers like process industry and the larger kind of distributed generation like wind turbines and windfarms.

- **Low Voltage grid**, or *distribution grid*. Voltage is below $1kV$, and its network structure is called *radial* because there are no rings and cycles in it. It is connected to the MV grid by so called *distribution stations*.

4.3.1.2 Power flow equations and power loss

The usual physical representation of a transmission line between bus i and j is:

That is, an ideal line (ij) containing a *resistance* r_{ij}, and a *reactance* x_{ij}. All equations here are for a line (i,j). The *impedance* of this line then, is $Z_{ij} = r_{ij} + ix_{ij}$[8]. This means that the relation between current and voltage in the line can be described by Ohm's law $I_{ij} = \frac{V_i - V_j}{Z_{ij}}$. Assuming a complex voltage of $V_i = v_i e^{i\theta_i}$ at bus i, and of $V_j = v_j e^{i\theta_j}$ at bus j, yields exercised powers of

$$S_{ij} = V_i I^* = V_i \frac{V_i^* - V_j^*}{Z^*} = \frac{v_i^2 - v_i v_j e^{i(\theta_i - \theta_j)}}{r - ix},$$

$$S_{ji} = \frac{v_j^2 - v_i v_j e^{i(\theta_j - \theta_i)}}{r - ix}.$$

Note that for ease of notation from here on r and x will be written instead of r_{ij} and x_{ij}, because the context makes clear which resistances and reactances are meant.

---

[8] capitals are used to represent complex quantities.

4.3 Optimal placing of wind turbines

Adding the complex powers exercised by voltages V_i and V_j at both ends of the line yields a total complex power conversion of

$$S_{tot} = S_{ij} + S_{ji} = \frac{v_i^2 + v_j^2 - v_i v_j (e^{i(\theta_i - \theta_j)} - e^{i(\theta_j - \theta_i)})}{r + ix}.$$

The magnitude $|S_{ij}|$ of a complex power is called the *apparent power*. Because of Joule's law $Q = I^2 \Omega t$, it is the current which determines the capacity of a transmission line (before it is damaged by heat production). Moreover for fixed voltage magnitude, the apparent power $|S_{ij}| = |V_i I^*| = |V_i||I|$ is proportional to the current on a line. Therefore in a power grid line capacities can be expressed in maximum apparent power.

But the apparent power says nothing about the power conversion in the line. That can be seen by taking the real parts of the above expressions, using $cos_{ij}(\theta) = \cos(\theta_i - \theta_j)$ and $sin_{ij}(\theta) = \sin(\theta_i - \theta_j)$, representing the *real powers* exercised by voltages V_i and V_j:

$$p_{ij} = \frac{1}{r^2 + x^2}[rv_i^2 - rv_i v_j \cos_{ij}(\theta) + xv_i v_j \sin_{ij}(\theta)],$$

$$p_{ji} = \frac{1}{r^2 + x^2}[rv_j^2 - rv_i v_j \cos_{ij}(\theta) - xv_i v_j \sin_{ij}(\theta)],$$

and adding them yields a total real power conversion of

$$p_{loss} = p_{ij} + p_{ji} = \frac{1}{r^2 + x^2}\left[r(v_i^2 + v_j^2) - 2rv_i v_j \cos_{ij}(\theta)\right].$$

This formula represents energy loss per second, depending on complex voltages on both ends of the line.

If we take the imaginary parts of the complex powers, we get the so called *reactive powers*

$$q_{ij} = \frac{1}{r^2 + x^2}[xv_i^2 - rv_i v_j \sin_{ij}(\theta) - xv_i v_j \cos_{ij}(\theta)],$$

$$q_{ji} = \frac{1}{r^2 + x^2}[xv_j^2 + rv_i v_j \sin_{ij}(\theta) - xv_i v_j \cos_{ij}(\theta)].$$

And these four equations describe the total (steady state) power grid behaviour, because Kirchhoff's Current Laws demand power flow conservation at every bus i for real power (that is, $\sum_j p_{ij} - \sum_j p_{ji} = 0$), and also for reactive power (that is, $\sum_j q_{ij} - \sum_j q_{ji} = 0$). By specifying certain boundary values at the buses (for example, p_{ij} and v_i at a generation bus i), and one reference node with phase $\theta = 0$, one can then solve for all voltage magnitudes and phases in the system. This is called AC power flow calculation.

4.3.1.3 DC load flow model

For technical purposes it is often required to be able to calculate the behaviour of bus voltage magnitudes and angles. It is important to maintain voltage stability, even when power demand and generation is subject to change. Unexpected reactive power behaviour may result in local voltage drops (which is undesirable for customers), or voltage rises (which may damage the electrical components of the system). Therefore the full solution of the above non-linear equations is critical for the technical operation of the system. However for general design questions and performance estimations, it is expedient to simplify a little further. Purchala et al. [125] justify the following assumptions for the context of this chapter:

1. The bus voltage magnitudes v_i and v_j are almost equal, so that we can say $v :\approx v_i \approx v_j$. That is, the effect of voltage drop on power system behaviour is neglected. Certainly for high voltage systems this is justified.

2. The difference in voltage angles $(\theta_i - \theta_j)$ is small. In this case one can make the first order approximations $\sin_{ij}(\theta) \approx (\theta_i - \theta_j)$ and $\cos_{ij}(\theta) \approx 1$.

3. An unnecessary assumption for our modelling purposes, but nevertheless true in practice for HV transmission lines, is $r << x$ (resistance is much smaller than reactance).

Making these approximations, and putting $v = 1$ (which is called expressing the voltage in per-unit) yields for our real power equations

$$p_{ij} = \frac{1}{r^2 + x^2}[rv_i^2 - rv_iv_j \cos_{ij}(\theta) + xv_iv_j \sin_{ij}(\theta)]$$

$$\approx \frac{1}{r^2 + x^2}[rv^2 - rvv + xv^2(\theta_i - \theta_j)]$$

$$= \frac{x}{r^2 + x^2}(\theta_i - \theta_j) \approx \frac{1}{x}(\theta_i - \theta_j),$$

$$p_{ji} \approx \frac{x}{r^2 + x^2}(\theta_j - \theta_i) \approx \frac{1}{x}(\theta_j - \theta_i).$$

One can see that due to $v_i = v_j$, the first and second term in square brackets cancel out. The justification of this approximation for HV power grid is even more enhanced by the third assumption that x tends to be (much) larger than r, making the third term dominant already.

Now compare a purely resistive DC electrical circuit where the only active elements are current sources connected to ground. In this case the current flow equations are $I_{ij} = \frac{1}{R_{ij}}(V_i - V_j)$ for all lines (ij).[9]

Note the similarity between the simplified power equations of the AC power grid, and the current equations of the DC electrical circuit. Their structure is exactly the same, only quantities differ. Thus it can be seen that after these simplifications a remarkable analogy exists between a large AC power system, and a DC electrical circuit equivalent. The role of currents in the DC load flow is taken over by powers in the AC network; the role of DC voltage drops over resistors correspond to differences between bus voltage angles, and DC resistance values correspond to reactance x of a transmission line.

It is according to this analogy that we like to think about the power grid as if there are power sources, power sinks, a power flow through the network, (even 'power conservation laws'), *en passant* producing line losses.

---

[9]here capitals represent normal real DC quantities.

4.3.1.4 Reactive power and line loss in the DC model

The same approximations 1) and 2) from the previous section, applied to the reactive power equations, yield:

$$\begin{aligned}
q_{ij} &= \frac{1}{r^2 + x^2}[xv_i^2 - xv_iv_j \cos_{ij}(\theta) - rv_iv_j \sin_{ij}(\theta)] \\
&\approx \frac{1}{r^2 + x^2}[xv^2 - xvv + xv^2(\theta_i - \theta_j)] = \frac{-r}{r^2 + x^2}(\theta_i - \theta_j), \\
q_{ji} &\approx \frac{-r}{r^2 + x^2}(\theta_j - \theta_i) = \frac{r}{x}p_{ij},
\end{aligned}$$

that is, if also $r \ll x$, reactive power 'flows' vanish, or at least are negligible in comparison to real power flows.

Similarly, the first order approximation for real power loss yields $p_{loss} = p_{ij} + p_{ji} = 0$. The DC model assumptions formulated at the beginning of the previous paragraph have as a direct consequence that zero power loss is assumed. This can easily be verified by the consideration that for all i and j, $v_i = v_j$, which means that there is zero voltage drop in the system, and consequently there can be no net power conversion. But taking the second order approximation $\cos(\theta) \approx 1 - \frac{\theta^2}{2}$ produces

$$\begin{aligned}
p_{loss} &= \frac{1}{r^2 + x^2}\left[r(v_i^2 + v_j^2) - 2rv_iv_j \cos_{ij}(\theta)\right] \\
&\approx \frac{1}{x^2 + r^2}\left[2r - 2r(1 - \frac{(\theta_i - \theta_j)^2}{2})\right] \\
&\approx rp_{ij}^2.
\end{aligned}$$

These powers do not actually disappear from the power flows in the model (which are, like the current in an electrical circuit, preserved), and therefore do not influence the power flow solution.

4.3.1.5 Network flow model

Above it was established by simplifying the exact AC power flow equations, that large scale AC power flow behaves approximately like a simple small scale DC electrical circuit. In former times this similarity was used by the old "DC network analyser", in which each network branch was represented by a resistance proportional to its series reactance and each DC current was proportional to a real power flow. The DC model derived its name from this analogue computing table.

These days computer algorithms can be exploited. By basic physics of electricity, Kirchhoff's Current Laws (KCL) and Kirchhoff's Voltage Laws (KVL) yield the equations by which the solution for the nodal voltages and currents over the lines can be computed. KCL demand that the total current into a node must equal the total current out of that node. KVL demand that the directed sum of the voltage drops over every closed loop in the network must equal zero. By simple linear algebra techniques the resulting system of linear equations can be solved. In power flow analysis, the matrix of coefficients of the linear equations is called the *admittance matrix*. The solution procedure comes down to inverting this matrix.

In the context of this chapter (stochastic and constrained optimization) another approach is more useful and also more intuitive. Crevier [31] remarks that the currents found by solving both Kirchhoff's Current and Voltage Laws, happen to be the same currents that minimize the total heat dissipation described by $\sum (I_{ij})^2 r_{ij}$, while satisfying only Kirchoff's Current Laws.

Because in the terminology of network flow optimization these current laws are just the flow balances at a node and because the objective function that describes total heat dissipation is clearly separable and convex (quadratic), we know from theory on network flows like Ahuja et al. [3] that the power flow problem can be solved efficiently by network flow algorithms. The advantage of this method over the traditional former one is that network algorithms allow for straightforward sensitivity analysis, can cope with additional constraints that may come in, and offer a known framework for a stochastic extension.

Concretely, this means that to find the solution for the currents through a DC network, and therefore an approximate solution for the power flow in an AC power system, the following mathematical program should be solved:

$$\min \sum_{(ij) \in A} p_{ij}^2 x_{ij},$$

subject to

$$\sum_{j:(ij) \in A} p_{ij} - \sum_{j:(ji) \in A} p_{ji} = b_i \quad (i \in N),$$

$$p_{ij} \geq 0 \quad ((ij) \in A),$$

where p_{ij} is the power flow through line (ij), x_{ij} is the reactance of the line, and b_i is the power supply/demand (the *balance*) of each node i.

From this it can be seen that the problem of computing the power flow solution of a power grid, is equivalent to solving a so called minimum cost flow problem with quadratic arc costs. Recall (e.g. from [3]) that the flow balance constraints may be written as $B\vec{p} = b$, where B is the so-called *node-arc incidence matrix* of the network, $\vec{p}$ is the vector of power flows through the arcs, and b is the vector of supplies at the nodes. Because of notational conventions, in the next chapters we will use y_{ij} instead of p_{ij} as symbol for the power flows.

4.3.1.6 Additional remarks

For the application in this chapter equal voltages are assumed throughout the grid. But when this is not the case, the above network flow program can easily be adjusted to account for the voltage differences. Define all arc cost coefficients

$$c_{ij} = \frac{x_{ij}}{v_{ij}},$$

where v_{ij} is the voltage at which the line is kept. The intuitive interpretation of this is, that the high voltage lines are 'cheaper' to travel over, and therefore are more likely

4.3 Optimal placing of wind turbines

to attract flow than lower voltage lines. This is nice since higher voltage lines typically cause less power loss, and have higher capacities.

The application of network algorithms to solve the DC power flow problem seems to be little advocated in literature. Yet modern day algorithms such as network simplex and ϵ-relaxation can solve quadratic cost flows extremely efficiently. It has been suggested that these methods are faster than the admittance matrix inversion of classical DC algorithms (e.g., [3], chapter 1). Moreover, strongly polynomial algorithms have been developed for separable quadratic cost *generalized* flows. In such network problems, a so-called *arc multiplier* is added to all arcs. The outgoing flow of the arc is then defined as the incoming flow times the arc multiplier. The QRELAXG algorithm, together with mathematical proof of its polynomiality, is presented in Tseng and Bertsekas [147].

This opens up the possibility to a DC algorithm where power losses are incorporated by a linear (first order) approximation of the losses over a branch. Compare Stott et al. [137] where a zero'th order approximation is suggested for the losses by subtracting fixed loss estimates from the node balances. There it is proposed (section VII-B) that incorporation of the losses could be achieved by estimating all powerflows p_{ij} beforehand by $\hat{p}_{ij}$. Then $r_{ij}\hat{p}_{ij}^2$ estimates the losses over this line. These losses can be subtracted from the node balances of the adjacent nodes before computing the power flows. According to [137], the losses "usually converge" when this process is iterated (using the resulting power flow as the new estimates $\hat{p}_{ij}$).

The quadratic cost generalized flow however, immediately yields losses that are proportional to the power flow over the branch. To determine the arc multipliers one should still estimate the power flows beforehand. Then, based on the estimate $\hat{p}_{ij}$, put the arc multiplier $\gamma_{ij} = r_{ij}\hat{p}_{ij}$ such that the linear loss $\gamma_{ij}p_{ij} = r_{ij}\hat{p}_{ij}p_{ij}$ approximates the actual loss $r_{ij}p_{ij}^2$. The network programming formulation looks like:

$$\min \sum_{(ij) \in A} p_{ij}^2 c_{ij},$$

subject to

$$\sum_{j:(ij)\in A} p_{ij} - \sum_{j:(ji)\in A} \gamma_{ji}p_{ji} = b_i \quad (i \in N),$$

$$p_{ij} \geq 0 \quad ((ij) \in A),$$

and take arc multipliers $\gamma_{ij} = r_{ij}\hat{p}_{ij}$. Iteration by using the resulting p_{ij} as the new flow estimates $\hat{p}_{ij}$, can be expected to converge both faster and more likely than the method of [137] because linear approximation of the quadratic loss function is significantly better than approximation by a constant function.

The model can be improved yet one step further. As will be shown in chapter 5, every quadratic network flow can be approximated to any desired degree of accuracy by a linear network flow in which parallel arcs are added. The flows over these parallel arcs sum up to the flow over the original quadratic cost arc. When decreasing arc multipliers are associated with these parallel arcs, a piecewise linear approximation of the quadratic loss can be realized. This is shown in Figure 4.8. Thus it can be seen that the network flow approximation to the DC power flow problem is flexible enough to realize the incorporation of quadratic losses into the DC power flow equations. We have seen this remark nowhere else in literature. It would be interesting to assess if this improvement makes the DC power flow model more suitable for MV grids, or very large scale HV systems, where power losses are more substantial.

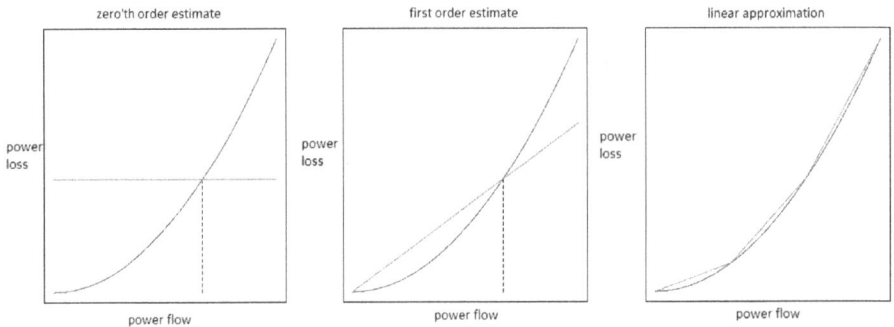

Figure 4.8: Relations of actual and estimated power loss on a line for the three methods. On the left: the method of [137]. In the middle: quadratic generalized network flow. On the right: linear generalized flow on the network with parallel arcs.

4.3.2 Problem definition

4.3.2.1 General outline

We expect HV and MV power grids to satisfy our DC load flow assumptions, and because the supply and demand of power have to be equal at every point in time, network behaviour of the HV/MV network can be modelled by solving a minimum cost flow. The node balances required by such a model can be obtained from the load data of the power network in the Netherlands. These data can be taken from a moment of peak load or of average load. The connection of a wind turbine to the grid corresponds to the increase of the supply at that particular node. In practice however, wind power production is highly uncertain. Although power demand by consumers also involves uncertainty, it is far better to predict than wind power production, and for that reason the assumption of deterministic demand seems justified. This assumption is also made by Carpinelli et al. [21].

In the figure below a rough sketch is provided, which may help to visualize the model described. In the original situation, power supply by the power plants (which are mainly connected to the upper parts of the grid) equals total demand by consumers (which are located mainly in the lower parts of the grid). The wind turbines to be placed provide an extra (stochastic) supply.

Of course, wind speed is geographically correlated: a strong wind in one place makes a strong wind in an other place, 100 km away, more likely. In order to model wind speed

4.3 Optimal placing of wind turbines

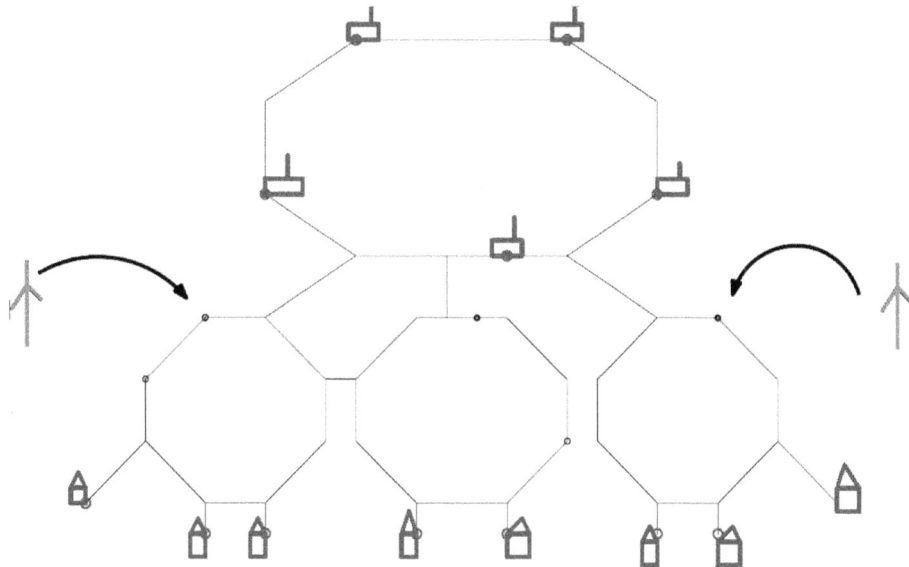

Figure 4.9: Schematic overview of placement problem.

distribution over the Netherlands, we shall use wind speed measurements throughout the country to partition the Netherlands into several zones. Within these zones we will assume wind speed to be 100% correlated, and we shall use the measurement data to estimate the correlations between the zones.

A few things can be observed:

1. Because power supply and demand always have to be equal, a choice has to be made what to do with the excess power in the system, resulting from the wind turbines we place. A few possibilities are: sell excess power abroad; increase demand of some demand node(s); decrease generation of some supply nodes. In this last case a choice has to be made which power plants to adjust. Because in practice market balancing happens non-locally, we propose to adjust for every MW of wind power, certain other power plants distributed over the country.

2. In general wind turbines are placed 'lower' on the grid than normal power plants. So as long as wind power production does not exceed local demand, wind turbines will have a positive effect on the transportation losses.

3. There is no intrinsic mechanism in the model that promotes placing the wind turbines on a spot with more wind than elsewhere (except when it can fulfil local demand). This might seem counterintuitive, but it is the consequence of our model. Models that minimize operating costs, do promote windy spots, because wind turbines have high building costs but low marginal costs. Therefore windy spots will make more profitable investments.

4. The model does promote spreading the wind turbines over different zones, but

only if peak wind power production is higher than local demand. In that case the peak spreading results in less power transport, and therefore less energy losses.

In this network model terminology, the research question translates to finding the optimal location of some wind turbines that have uncertain but correlated supply (depending on the geographical zone), such that the expected transportation losses are minimized. In order to maintain global balance, we adjust for the supply of these wind turbines, by decreasing the supply of the original supply nodes.

In the following two sections, a step-by-step procedure is described to capture the model and the research question in a mathematical formulation. First the model is described for computing the power flow and line losses, once all power productions are given. This may be called the second stage of the problem, because wind turbines have to be placed first. Next the model is described for the placement of the wind turbines, and for computing the produced wind power and node balance corrections. This may be called the first stage of the problem. Note that the second stage will be described prior to the first stage. The two stages and the stochasticity involved are represented intuitively in Figure 4.10.

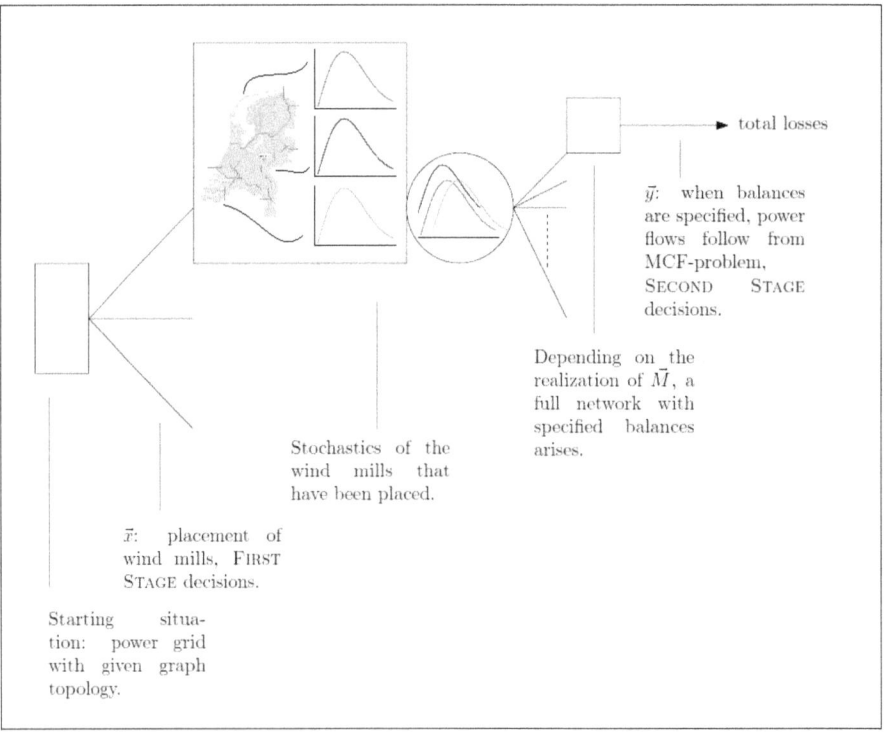

Figure 4.10: Illustration of the modelling steps.

4.3.2.2 Power flow problem

Define an (undirected) graph $G = (N, A)$, taking a node $n \in N$ for every bus in the power grid, and an arc $a \in A$ for every transmission line connecting two buses. Define a cost function $c : A \to \mathbb{R}^{\geq 0}$, which associates with every arc $a_{ij} \in A$ a cost c_{ij}, and take a cost c_{ij} equal to the reactance X_{ij} of its corresponding transmission line. Define a balance function $b : N \to \mathbb{R}$, associating with every node i its supply of power b_i: a negative number for load buses, a positive number for generation buses. Define a nonnegative power flow $y : A \to \mathbb{N}$ on the arcs of G, satisfying the balance equation $\sum_{j:(ij)\in A} y_{ij} - \sum_{j:(ji)\in A} y_{ji} = b_i$ for each node i, and $y_{ij} \geq 0$ for all $(ij) \in A$. Assuming a power balance $\sum_{i \in N} b_i = 0$, the so called *power flow solution* is found by minimizing $\sum_{(ij)\in A} c_{ij} y_{ij}^2$ over all feasible power flows.[10] Once the power flows have been found, $\sum_{(ij)\in A} y_{ij}^2 r_{ij}$ is the expression for the total power loss.

4.3.2.3 Wind turbine placement problem

Partition the node set N into k zones such that $N = N_1 \cup N_2 \cup \cdots \cup N_k$. Introduce the random vector $\widetilde{M} \in \mathbb{R}^k_{\geq 0}$, representing the uncertain wind speeds in the zones at some arbitrary time instance; its k-th element $\widetilde{M}_k$ being the wind speed in zone k. Transform $\widetilde{M} \in \mathbb{R}^k_{\geq 0}$ into $M \in \mathbb{R}^n_{\geq 0}$ by firstly, computing for each zone $i = 1, \ldots, k$ what wind power is produced by one wind turbine from the wind speed in zone i, and secondly, copying this number for each node in that zone. That is, if two nodes i, j are in the same zone, then $M_i = M_j$. We call M_i the *potential wind speed* at node i. Introduce decision variables $x_i : i \in N$, representing the choice of placement for the wind turbine in zone i. Therefore for all regular $i : x_i \in \{0, 1, \ldots\}$ and $\sum_i x_i = l$ should be satisfied, where l is the number of wind turbines to be placed.[11] Now $\sum_{i \in N} M_i x_i = (\vec{M} \cdot \vec{x})$ represents the total wind power produced, so if $t \in \mathbb{R}^n$ represents the fractions of production adjustments to compensate for the wind power (with $\sum_{i \in N} t_i = 1$), then $b_i + M_i x_i - (\vec{M} \cdot \vec{x}) t_i$ is the new balance for any node $i \in N$. Note that $\sum_{i \in N} \left[b_i + M_i x_i - (\vec{M} \cdot \vec{x}) t_i \right] = (\sum_{i \in N} b_i) + (\sum_{i \in N} M_i x_i) - \sum_{i \in N} (\vec{M} \cdot \vec{x}) t_i = 0 + \vec{M} \cdot \vec{x} + (\vec{M} \cdot \vec{x}) \sum_{i \in N} t_i = 0$, so total power balance is indeed always preserved, no matter what the value of $\vec{M}$ may happen to be. Then given a placement vector $\vec{x}$ and a realisation $\vec{m}$ of $\vec{M}$, the vector of power flows $\vec{y}$ can be computed as follows:

$$\vec{y} = \text{argmin} \sum_{(ij)\in A} c_{ij} y_{ij}^2,$$

---

[10] Note that it is necessary that the balances sum up to zero, or, total power production equals total power consumption. This is because the DC power flow model assumes that no power losses occur. Power losses are neglected while computing the power flow solution. Afterwards, when the power flow solution has been computed, an estimate is made of the losses that have occurred.

[11] Do not confuse this decision variable x_i with the symbol x_{ij} or x, used in the previous sections to denote line reactance. Notational conventions in power engineering demand the use of x_{ij} for reactance, and in mathematical programming it is customary to denote first stage decision variables by x_i.

subject to
$$y_{ij} \geq 0 \quad ((ij) \in A),$$
$$\sum_{j:(ij) \in A} y_{ij} - \sum_{j:(ji) \in A} y_{ij} = b_i + m_i x_i - (\vec{m} \cdot \vec{x}) t_i \quad (i \in N).$$

It is clear that the power flows y_{ij} depend on the placement decisions x_i and on the random vector M, which determine the balances of the minimum cost flow. Hence, different choices of x_i will mount to different values of y_{ij}, and therefore different transportation losses. Moreover, the expected value of $y_{ij}(\vec{M}, \vec{x})$ over $\vec{M}$ is well defined. The aim then to minimize expected transportation losses by optimally placing the wind turbines may be written in the following way:

$$\min \mathbb{E}_M \left[\sum_{(ij) \in A} y_{ij}^2 r_{ij} \right],$$

subject to
$$x_i \in \{0, 1, \dots\} \quad (i \in N),$$
$$\sum_{i \in N} x_i = l.$$

4.3.2.4 The complete model

Combining the two previous sections, a complete formulation is provided by

$$\min \mathbb{E}_M \left[\sum_{(ij) \in A} y_{ij}^2 r_{ij} \right],$$

subject to
$$x_i \in \{0, 1, \dots\} \quad (i \in N),$$
$$\sum_{i \in N} x_i = l,$$

and where
$$\vec{y} = \operatorname{argmin} \sum_{(ij) \in A} c_{ij} y_{ij}^2,$$

subject to
$$y_{ij} \geq 0 \quad ((ij) \in A),$$
$$(By)_i = b_i + m_i x_i - (\vec{m} \cdot \vec{x}) t_i \quad (i \in N).$$

In the last equation (which represents the flow balance constraints), B is the node-arc incidence matrix for G. The right-hand side $b_i + m_i x_i - (\vec{m} \cdot \vec{x}) t_i$ are the balances, adjusted for the placement of the wind turbines.

4.3.3 Solution method

In Section 4.3.2.4 a complete mathematical formulation of the research question was given. In this section we try to reformulate the problem into a standard solvable minimization problem. First, the stochastic program is rewritten into a large deterministic one by introducing scenario parameters. These parameters will be obtained by simulation. Next, a linearization of the network flow that arises in the second stage is described. Then it is shown how duality theory of linear network flows may be used to combine the two stages into one large optimization problem. Finally, we present the solution strategies for this problem.

4.3.3.1 Stochastic programming

Models like the one above are called two stage stochastic optimization problems:

- First stage: the decisions $\vec{x}$ have to be made while the future behaviour of M is still uncertain; then a realization of M is observed;

- Second stage the laws of power electronics determine the decision variables $\vec{y}$, which results in a value of the objective function, that is, the transportation losses.

The first difficulty in carrying out the minimization above, has to do with the form of the objective function. Deterministic optimization algorithms cannot trivially handle an objective function like

$$\mathbb{E}_M \left[\sum_{(ij)\in A} \left(y_{ij}(\vec{M}, \vec{x}) \right)^2 r_{ij} \right],$$

which contains an expectation. (The notation $y_{ij}(\vec{M}, \vec{x})$ makes the dependence clear of the y_{ij} on $\vec{M}$ and $\vec{x}$.) The random variable $\vec{M}$ has a continuous (multidimensional) distribution, and it is difficult to see how a (multidimensional) integral in the objective can be dealt with. Next to this, an other possibility is solve this as a Stochastic Recourse Model.. However, then we need both a linearization of the objective function and a relaxation of the integer decision variables.

Every continuous distribution can be approximated by a discrete one, and moreover, one which assumes finitely many values. Therefore also our wind speed distribution may be assumed discrete. And when it is assumed that $\vec{M}$ takes only finitely many different values, the expectation can be computed as a finite sum. To this end, define a set S containing finitely many so-called *scenarios* $s \in S$. Next assume that $\vec{M}$ will take the value $\vec{m}^s$ with probability p^s, where $\sum_{s \in S} p^s = 1$. Then the expectation of a function $f(\vec{y})$ of $\vec{y}$ can be written $\sum_{s \in S} p^s \cdot f(\vec{y}^s)$.[12]

In this way the stochastic program of the previous section can be written as an equivalent (large-scale) deterministic extension by summing over the scenario space:

---

[12]The scenario set S is a kind of finite discretization of the sample space Ω of M: in fact, the expectation in the objective function could have been written $\int_{\omega \in \Omega_M} \cdot \left(\sum_{(ij)\in A} (y_{ij}^\omega)^2 r_{ij} \right) d\omega$, whereas the approximation by introducing the scenario set S looks like $\sum_{s \in S} p^s \cdot \left(\sum_{(ij)\in A} (y_{ij}^s)^2 r_{ij} \right)$.

$$\min \sum_{s \in S} p^s \cdot \left(\sum_{(ij) \in A} (y_{ij}^s)^2 r_{ij} \right),$$

subject to

$$x_i \in \{0, 1, \dots\} \quad (i \in N),$$

$$\sum_{i \in N} x_i = l,$$

and where

$$\vec{y^s} = \operatorname{argmin} \sum_{(ij) \in A} c_{ij}(y_{ij}^s)^2,$$

subject to

$$y_{ij}^s \geq 0 \quad ((ij) \in A, s \in S),$$

$$(B\vec{y^s})_i = b_i + m_i^s x_i - (\vec{m^s} \cdot \vec{x}) t_i \quad (i \in N, s \in S).$$

4.3.3.2 Linearization

Still this formulation does not fit a standard mathematical programming model. The first stage decisions $\vec{x}$ aim to minimize the expectation of some function of the second stage decisions $\vec{y^s}$, whereas these themselves must be chosen in such a way that they minimize some other function. At first sight it is not clear how to formulate the entire problem as a single optimization model. The constraint that the $\vec{y^s}$ are the arguments which solve another minimization problem, is quite an unusual one. But by a few steps this constraint may be written in a standard form which can be handled by optimization software. The first of these steps is the linearization of the objective functions.

Take the second stage minimization problem

$$\vec{y^s} = \operatorname{argmin} \sum_{(ij) \in A} c_{ij}(y_{ij}^s)^2,$$

subject to

$$y_{ij}^s \geq 0 \quad ((ij) \in A, s \in S),$$

$$(B\vec{y^s})_i = b_i + m_i^s x_i - (\vec{m^s} \cdot \vec{x}) t_i \quad (i \in N, s \in S).$$

As already observed, these second stage decisions solve $\#S$ independent but very similar convex network flows. (The only difference between scenarios is a smaller or larger change in supply of the wind turbines with their corresponding corrections.) The

4.3 Optimal placing of wind turbines

similarity between the scenarios may be used to obtain the solution of all $\#S-1$ network flows efficiently by starting out from the solution to the first one. Of convex network flows it is known that they can be approximated by linear network flows to any desired degree of accuracy by adding multiple arcs between the connected nodes, and defining appropriate arc cost and capacity parameters (as, for example, [3] do). This can be understood as follows.

Suppose some arc $(ij) \in A$ has capacity u_{ij} and cost function $c_{ij}(y_{ij})^2$, and suppose the cost function is approximated by a piecewise linear function consisting of H line segments as in Figure 4.11:

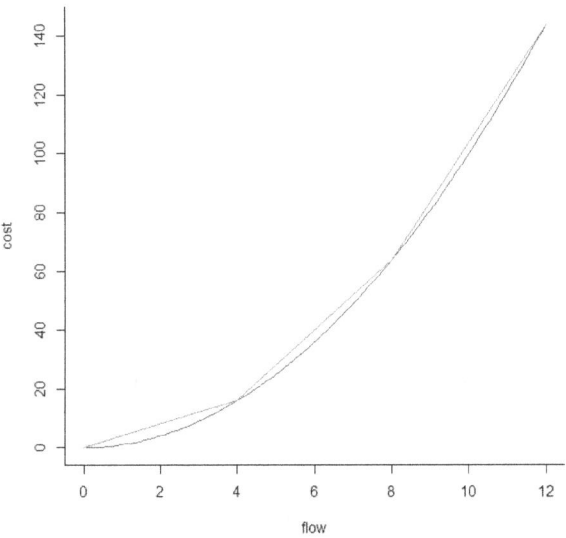

Figure 4.11: Linearization of the cost function.

Now this arc (ij) and the piecewise linear cost function on it behave in the same way as H parallel arcs (ij, h), $h = 1, \ldots, H$ with capacity $U_{ij,h} = \frac{u_{ij}}{H}$ and linear cost functions $C_{ij,h}$ that have a slope equal to the slope of the h'th line segment in the picture. This follows because a flow from i to j that should minimize the cost, will always first fill the cheapest arcs (ij, h), that is, from $h = 1$ upwards to $h = H$. The old cost of the flow over arc (ij) now equals the sum of the new flows over the arcs (ij, h).

By applying this transformation the number of nodes remains the same, and node balance constraints change only in the sense that there is one more arc parameter to sum over. The objective function becomes

$$\vec{y}^s = \operatorname{argmin} \sum_{(ij) \in A} \sum_{h \in \{1,\ldots,H\}} C_{ij,h} y_{ij,h}.$$

The value of the first stage objective function $(y_{ij}^s)^2 r_{ij}$ can easily be computed in terms of the new flow variables $y_{ij,h}$ and in a similar way newly defined $R_{ij,h}$. The total linearized problem then looks like:

$$\min \sum_{s \in S} p^s \cdot \left(\sum_{(ij) \in A} \sum_{h \in \{1,\ldots,H\}} R_{ij,h} y_{ij,h}^s \right),$$

subject to

$$x_i \in \{0, 1, \ldots\} \quad (i \in N),$$

$$\sum_{i \in N} x_i = l,$$

and where

$$\vec{y^s} = \operatorname{argmin} \sum_{(ij \in A)} \sum_{h \in \{1,\ldots,H\}} C_{ij,h} y_{ij,h}^s,$$

subject to

$$0 \leq y_{ij,h}^s \leq U_{ij,h} \quad ((ij, h) \in A^*, s \in S),$$

$$\sum_{j,h:(ij,h) \in A^*} y_{ij,h}^s - \sum_{j,h:(ji,h) \in A^*} y_{ij,h}^s = b_i + m_i^s x_i - (\vec{m^s} \cdot \vec{x}) t_i \quad (i \in N, s \in S),$$

where A^* denotes the new (strongly enlarged) set of arcs with linear cost, indexed by (ij, h), where $(ij) \in A, h \in \{1, \ldots, H\}$. Again the last equation represents the node balance constraints. By the introduction of the parallel arcs, each node now has more arcs incident to it than before, but the structure of the graph remains the same. If B^* denotes the node-arc incidence matrix for the new set of arcs, the flow balance constraints could have been written $(B^* \vec{y^s})_i = b_i + m_i^s x_i - (\vec{m^s} \cdot \vec{x}) t_i$ for $i \in N, s \in S$.

4.3.3.3 Reformulating the second stage constraint

The second and final step of rewriting the problem is performed in this section. When this has been done the problem of minimizing the transportation losses will be formulated in the form of a standard mathematical program.

The point of linearizing the network flow is that for linear network flows a well-known duality framework exists, in which *strong duality* holds: solving the minimum cost flow under flow nonnegativity and balance constraints is equivalent to (and yields the same optimal objective value as) solving another, related, linear program, called its *dual*. Bertsimas and Tsitsiklis [13] give a very nice introduction into general duality in section 4, and apply it to network flows in section 7. Alternatively [3], chapter 9 derive the dual network problem from first principles.

The general results may be summarized by the following. Let $z(y^*)$ denote the cost value of a minimum cost flow with flow variables $\vec{y}$ and balance vector $\vec{b}$. Define $\pi : N \to \mathbb{R}$, called the *node potential* which associates with every node i a real number π_i.

4.3 Optimal placing of wind turbines

Then maximizing $w(\pi) = \sum_{i \in N} \tilde{b}_i \pi_i - \sum_{(ij) \in A} \max(0, \pi_i - \pi_j - c_{ij}) u_{ij}$ yields an optimal node potential π^*, and $w(\pi^*) = z(y^*)$. Therefore, when faced with a linear network flow problem, it can be checked if it has been solved to optimality by a simple condition: does a node potential π exist for which $w(\pi)$ equals the current flow cost?

This theory can be applied in the following way. In the two stage model above, one of the constraints is that the $\vec{y}^s$ have to be chosen such that they minimize some flow cost objective $\sum_{ij,h \in A^*} C_{ij,h} y^s_{ij,h}$ called the *primal value*. For the optimal $\vec{y}^s$, there exist node potentials $\vec{\pi}^s$ for which the *dual value* $w(\pi^s)$ equals the primal. Hence in stead of the 'argmin'-constraint, it is equivalent to add free node potential variables for every scenario, and add the constraint that for every scenario primal value should equal dual value.

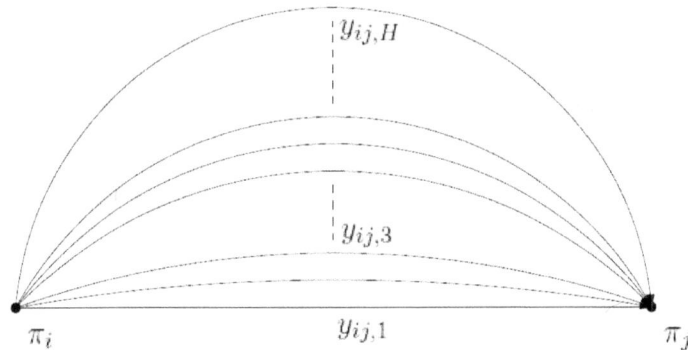

Figure 4.12: Node potentials and flow variables for the parallel arcs.

Let $\tilde{b}^s_i = b_i + m^s_i x_i - (\vec{m}^s \cdot \vec{x}) t_i$ be the balance at node i, corrected for scenario s and the placement decisions $\vec{x}$. Then the condition that $\vec{y}^s = \operatorname{argmin} \sum_{ij,h \in A^*} C_{ij,h} y_{ij,h}$, where $\vec{y}^s$ is a feasible flow, is equivalent to the condition that $\sum_{i \in N} \tilde{b}^s_i \pi^s_i - \sum_{(ij,h) \in A^*} \max(0, \pi^s_i - \pi^s_j - C_{ij,h}) U_{ij,h} = \sum_{ij,h \in A^*} C_{ij,h} y_{ij,h}$, where $\vec{y}^s$ is a feasible flow and $\vec{\pi}^s$ a node potential vector.

After these new variables are introduced and the constraint changed as described above, the model can finally be formulated as a standard optimization problem:

$$\min \sum_{s \in S} p^s \cdot \left(\sum_{ij,h \in A^*} R_{ij,h} y^s_{ij,h} \right),$$

subject to

$$x_i \in \{0, 1, \ldots\} \quad (i \in N),$$

$$\sum_{i \in N} x_i = l,$$

$$0 \leq y_{ij,h}^s \leq U_{ij,h} \qquad ((ij,h) \in A^*, s \in S),$$
$$\pi_i^s \in \mathbb{R} \qquad (i \in N, s \in S),$$
$$(B^* \vec{y^s})_i = b_i + m_i^s x_i - (\vec{m^s} \cdot \vec{x}) t_i \qquad (i \in N, s \in S),$$
$$\sum_{i \in N} \tilde{b}_i^s \pi_i^s - \sum_{(ij,h) \in A^*} \max(0, \pi_i^s - \pi_j^s - C_{ij,h}) U_{ij,h} = \sum_{ij,h \in A^*} C_{ij,h} y_{ij,h}^s \qquad (s \in S),$$

where $\tilde{b}_i^s$ denotes the corrected node balances, and B^* the node-arc incidence matrix adjusted for the parallel arcs.

4.3.3.4 Solution strategies

With respect to the solution of the mathematical program formulated in the last section, there are a few things that should be kept in mind. By the introduction of the new indices s for stochastic programming and h for linearization of the convex cost arcs, and the introduction of the new variables π^s, the whole program quickly becomes giant. For example, in the next section, 729 scenarios are defined, that is, $\#S = 729$, and if a transmission line of heat capacity $1000 MVA$ is to be modelled with precision $1MW$, then $H = 1000$. The result is, that the number of flow variables is multiplied by almost a million, and that in addition $\#S \cdot n$ new (node potential) variables enter the problem.

But there is one more (somewhat hidden) drawback of the method described in this section. The constraint that the dual must equal the primal value of the second stage was written

$$\sum_{i \in N} \tilde{b}_i^s \pi_i^s - \sum_{(ij,h) \in A^*} \max(0, \pi_i^s - \pi_j^s - C_{ij,h}) U_{ij,h} = \sum_{ij,h \in A^*} C_{ij,h} y_{ij,h}^s, \qquad (s \in S).$$

But the balance $\tilde{b}_i^s$ used in this equation crucially depends on the first stage decisions $\vec{x}$. It was defined $\tilde{b}_i^s = b_i + m_i^s x_i - (\vec{m^s} \cdot \vec{x}) t_i$, and therefore the product $\tilde{b}_i^s \pi_i^s$ contains a term $x_i \pi_i^s$; which makes the model nonlinear. Naive rigorous solution when the model is defined on any realistic scale (which may include hundreds of nodes and arcs, as well as hundreds of scenarios and parallel arcs) will probably be infeasible. Specialized algorithms may be designed by exploiting the primal-dual relationship between variables $y_{ij,h}^s$ and π_i^s, but are beyond the scope of this thesis.

Therefore in this chapter we present a heuristic approach, based on the observation that the functions $\sum_{ij} r_{ij} p_{ij}^2$ and $\sum_{ij} x_{ij} p_{ij}^2$ have the same structure. If for all branches ij, the fraction $\frac{x_{ij}}{r_{ij}}$ is the same, then minimization of $\sum_{ij} x_{ij} p_{ij}^2$ would immediately yield a placement that minimizes losses. Now minimize objective function $\sum_{ij} c_{ij} p_{ij}^2$, where the cost parameters c_{ij} are a mixture of reactance x_{ij} and resistance r_{ij}. Minimization of $\sum_{ij} x_{ij} p_{ij}^2$ is needed for correct power flows; minimization of $\sum_{ij} r_{ij} p_{ij}^2$ yields loss minimizing objective. So a certain tradeoff has to be made which aspect is more important for an arc. One simple suggestion could be to take cost parameters $c_{ij} = \frac{1}{2}(r_{ij} + x_{ij})$; or another

$$c_{ij'} = r_{ij'} \cdot \left(\frac{1}{m} \sum_{ij} \frac{r_{ij}}{x_{ij}} \right)^{-1}.$$

4.3 Optimal placing of wind turbines

Now the second stage of the problem disappears, which makes the problem much easier. Two extreme parameter choices can be identified. First: disregard resistances completely (that is, take $c_{ij} = x_{ij}$), which respects power flows, but disregards losses[13]. And second: disregard reactances completely (that is, take $c_{ij} = r_{ij}$), which does minimize losses, but does not respect power flows. Chen [27] makes this second choice (he does not respect power flows; only flow balances). It is difficult to say something intuitive about the quality performance of this heuristic; in general it will perform poorer if $\frac{x_{ij}}{r_{ij}}$ differs much for different arcs.

4.3.4 Wind power simulation

We performed a simulation of wind speeds in the Netherlands, based on empirical data and on wind speed models from literature. The goal is to compute reasonable values for the model parameters p^s and m^s that were used in the optimization model. Recall that the interpretation of p^s is the probability of a scenario $s \in S$, whereas m^s is a vector containing for scenario s the joint wind power output if on every node there would be exactly one wind turbine.

For this chapter data were taken from the hourly wind speed measurements of all 50 measurement stations during the years 2001 until 2010 from The Dutch meteorological institute (www.knmi.nl).

First we made some partition of the Netherlands into a few different wind zones, within which we will assume equal wind speeds based on *clustering* techniques, exact description can be found in Johnson and Wichern [74].

Next we modelled the wind speed at a particular site. In literature, the Weibull or even Rayleigh (which is a special case of the Weibull) distribution is advocated. See, for example, [46], or [21]. Conclusion of our analysis is that the assumption of Rayleigh distributed marginals is sufficiently justified.

Next we looked at the joint distribution of the wind speeds in the several regions. [46] suggest a multivariate distribution which is easy to simulate, and has marginals with approximate Rayleigh properties. They cite [73] and [72] as mathematically more precise, but of a (for this purpose) unjustified complexity.

Having simulated 10000 independent instances of joint wind speeds, these should now be transformed into the powers that are produced by a wind turbine within these zones. Four main characteristics are mentioned in literature: a wind speed value v_{ci} (cut-in velocity) below which the turbine has no output, a rated velocity v_r above which the turbine has its maximum output, and a cut-out velocity v_{co} above which the turbine is deactivated in order to prevent damage. The fourth characteristic is the rated or maximum power P_r itself.

Chen [27] takes as power output curve the left graph in Figure 4.13, which is doubtless the more accurate model. But we think that the linear approximation used by [21] and [114] suffices for our purposes.

There are many different sizes and types of wind turbines available in industry. For this chapter typical characteristics for a modern large $2500kW$ turbine were taken.

---

[13]This is the choice made to produce the results in the next section.

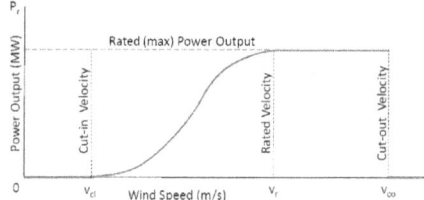 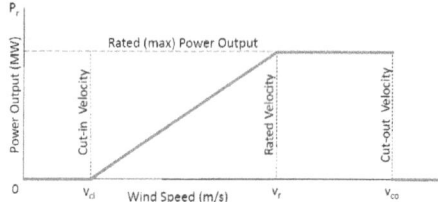

Figure 4.13: Non-linear and linear power curves.

By the procedure described until here, it is possible to simulate correctly correlated wind power outputs for single wind plants over the different regions. Total wind power output in a region can then be obtained by multiplying that value by the number of turbines in the region, as described in the previous section. This simulation procedure can be used to compute values for the scenario parameters p^s and m^s, defined for the stochastic optimization program.

p^s, the probability of a scenario $s \in S$, could be computed by partitioning the 6-dimensional range of power outputs, and counting which parts of the simulated values would lie in the particular areas. m^s, the value of the scenario, is then obtained by averaging over these values only. But the fact that we started the simulation from independent uniforms, provides us with an opportunity for simplification: by simply partitioning the range of the multivariate uniforms (that is, $[0,1]^6$), then mapping them to output powers and compute the m^s, it becomes easy to define scenarios which have all equal probability p^s.

Now that these parameters p^s and m^s, $s \in S$ have been computed, the deterministic optimization problem that was drawn up in the previous section can be solved.

4.3.5 Results

In this section the results of the performed computations are presented. In order to visualize the trade-offs mentioned in Section 4.1.1, first some preliminary results are shown: the trade-off between high expected value and small variance of the total wind power, is illustrated in Section 4.3.5.1 and the trade-off between local and central placement for stochastic supply, is illustrated in Section 4.3.5.2. In the last subsection the results of the full model are shown.

4.3.5.1 Significance of wind zones

In Figure 4.14, normalized output power histograms are shown for 6 wind turbines. The left picture applies to the situation where all outputs are 100% correlated, i.e., all turbines are in a single zone. The region of Groningen and Friesland is taken as an example, because it has medium expected wind speed. The right picture applies when the turbines are equally distributed over 6 different wind zones. The results are obtained from the data used and discussed in Section 4.3.4.

4.3 Optimal placing of wind turbines

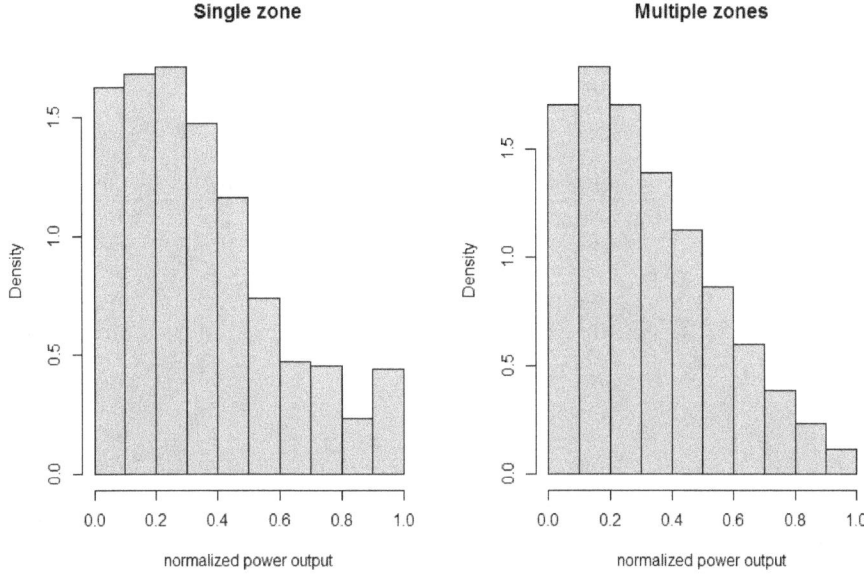

Figure 4.14: Histograms for power output in single and multiple zones.

Although the effect is not very strong, one can see that the distribution for the single zone is much flatter than the distribution for multiple zones. In fact, the variance of the single zone output is 1.5 times higher than for multiple zones. This suggests that there could be benefits from spreading wind turbines over the different wind zones. Of course the benefits in larger countries would be greater (for example, [27] presents similar histograms for various sites in the United States, indicating a stronger effect in variance reduction).

Now suppose that we distribute 20 wind turbines over the various zones, yielding a placement vector p which contains for every zone $i = 1, \ldots, 6$ the number of wind turbines p_i that are allocated to that zone. Define the output of a single wind turbine in zone i as X_i, $i = 1, \ldots, 6$ and define the total power output of the 20 turbines $X_p = \sum_{i=1}^{6} p_i X_i$.

For various reasons a high variance of the output could be undesirable. In our context high variance of power output would mean that it is more difficult to match it to local demand, resulting in higher power losses by transportation. In finance and trade, high variance of production means less predictable volumes to be traded, resulting in lower selling prices. In order to compare expected value and variance, it has been proposed (for example Steinbach [136]) to maximize the function $f(p) = E[X_p] - \lambda Var(X_p)$, where λ represents the costs of volatility. This so-called *mean-variance model* is common in financial optimization and other disciplines. The variance of X_p can be expressed as

$$var(X_p) = var(\sum_{i=1}^{6} p_i X_i) = \sum_{i=1}^{6}\sum_{j=1}^{6} cov(p_i X_i, p_j X_j) = \sum_{i=1}^{6}\sum_{j=1}^{6} p_i p_j cov(X_i, X_j),$$

where the covariance matrix Σ of the X_i is known from the wind data discussed in Section 4.3.4. Let $\mu_i = E[X_i]$, and let $\sigma_{i,j} = cov(X_i, X_j)$ be fixed parameters. Then the function $f(p)$ can be written in the nonlinear form

$$f(p) = \sum_{i=1}^{6} p_i \mu_i - \lambda \sum_{i=1}^{6}\sum_{j=1}^{6} p_i p_j \cdot \sigma_{i,j},$$

and its variables p_i, $i = 1,\ldots,6$ are subject to the constraints $p_i \in \{0,\ldots,20\}$, and $\sum_{i=1}^{6} p_i = 20$. The parameters are given by

$$\mu = \begin{pmatrix} 65.77 \\ 56.95 \\ 37.00 \\ 41.74 \\ 41.18 \\ 50.47 \end{pmatrix}, \quad \Sigma = \begin{pmatrix} 1008.7 & 699.09 & 480.54 & 542.84 & 585.35 & 538.84 \\ 699.09 & 781.24 & 415.07 & 438.21 & 455.23 & 569.42 \\ 480.54 & 415.07 & 389.71 & 381.42 & 394.28 & 377.39 \\ 542.84 & 438.21 & 381.42 & 526.13 & 479.52 & 409.67 \\ 585.35 & 455.23 & 394.28 & 479.52 & 591.54 & 412.97 \\ 538.84 & 569.41 & 377.39 & 409.67 & 412.97 & 678.27 \end{pmatrix}.$$

Again AIMMS can solve this for various λ. The problem is a Mixed Integer Quadratic Programming (MIQP) problem with 6 variables. Table 4.4 shows results for various values of λ. Because of the high values in matrix Σ compared to μ, the range of λ for which interesting changes occur, is quite near to zero ($\lambda \in [0, 0.01]$).

100λ	0	0.1	0.2	0.3	0.4	0.5	0.6	0.7	1
North Sea Coast	20	15	9	4	2	0	0	0	0
Zuiderzee	0	4	4	3	3	2	2	1	0
Inlands	0	0	0	6	9	11	13	14	15
Limburg	0	0	0	2	2	3	2	2	2
WN Brabant	0	0	0	0	0	0	0	0	0
Groningen Friesland	0	1	7	5	4	4	3	3	3

Table 4.4: Distribution of turbines for various λ in the mean variance model.

For $\lambda = 0$, naturally all turbines are placed in the zone with highest average wind speed. For λ too high, the term $-\lambda Var(X_p)$ becomes dominant, and most turbines are allocated to the zone 'Inlands'. This zone has lowest variance, but it has also low expected value.

The value of λ that should be taken for practical optimization, is determined by external factors (market volatility, share of wind power in the generation portfolio of the power company). One can clearly observe that the optimal wind turbine distribution for a power company strongly depends on the value of its λ.

4.3.5.2 Network structure

In the former section only wind speed behaviour was taken into account, while network structure of the power grid was neglected. An optimal dispersion over the various wind zones was achieved, depending on the relative importance of expectation and variance of the output. In this section, by contrast, the structure of the power grid is analysed. This is done by placing DG units in the network, and assessing their effect on the total line losses on the grid. Optimal locations for both deterministic and stochastic DG units are presented. The problem is similar to the placement of wind turbines, but simpler because wind speed correlation structure is disregarded.

As a first orientation into the structure and the load of the HV grid, the simple power flow problem can be solved. It yields the following observations:

- The global structure of the grid resembles two connected rings, as depicted in the left part of Figure 4.15. The province of Groningen and the loose ends in the west are exceptions to this, but the major part of the transportation is taking place at one cycle (mainly from West to East).

- For achieving minimum transportation losses, DG units are very likely placed on nodes with negative supply. Therefore if node 619 (Diemen), which has positive supply, is moved to the wind zone 'North Sea Coast', this is likely to have no consequences for any solutions.

- Almost all nodes in 'North Sea Coast' and in 'West Noord-Brabant' are supply nodes.[14] It is unattractive to place DG units there. When these two zones are united, and the rest of the nodes are sorted according to their wind zones, a picture can be made of the power exchanges between the several zones. This is shown in the right part of Figure 4.15. Net supplies are computed per zone, and the power exchange between the various zones is the sum over one or two transmission lines. Both quantities are in MW.

- From these pictures it is intuitively clear that DG units will be placed most likely in the 'Inlands' or 'Limburg'.

- Total load on the grid is $13,325$MW. About half of it is exchanged between the zones. The network losses for the standard power flow solution amount to 7.5MW.

In order to investigate network preferences for the placement of DG, Tables 4.5 and 4.6 are presented. They contain the results of a sequence of placement problems for varying DG penetration, and for varying size of the generation units. By penetration is meant: total added generation capacity divided by total original generation capacity ($13,325$MW).

---

[14] The large supply at Beverwijk is explained by offshore wind farms at the height of Noord Holland. The large supplies at Maasvlakte and Geertruidenberg are explained by the large number of Combined Heat and Power (CHP) systems used by greenhouses and industry. Borssele is home to a nuclear power plant.

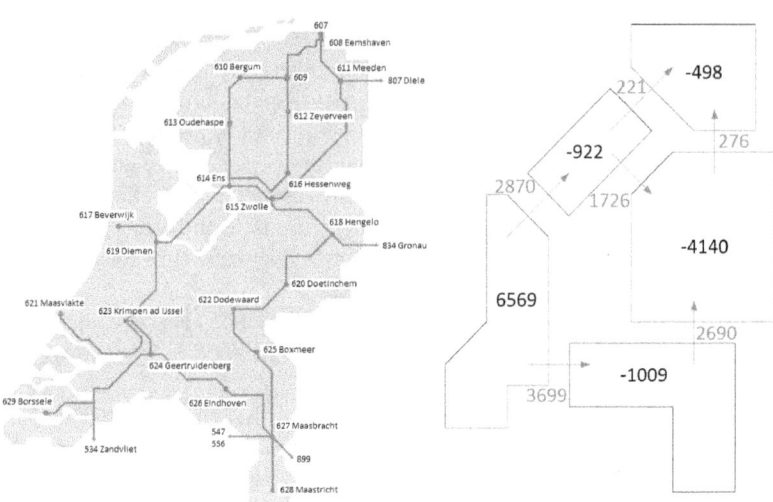

Figure 4.15: Power exchanges in MW between the wind zones for the standard power flow solution.

In the limit of 100% penetration and small enough unit size, it is clear that transportation losses can be made to vanish. When distributed generation equals demand at every node, no power transportation takes place at all.

The extreme case of placing only one DG unit can be used to obtain the notion of a kind of centrality of a node. When placing one unit of 1MW, the location which minimizes the loss is the point for which the network losses are most sensitive to a change in demand. One DG unit with 100% penetration yields the best location if the whole network would be powered from one source node.[15] The resulting source node turns out to be Doetinchem for small size, Dodewaard when a size of $6,000 MW$ is reached, and Boxmeer when 100% penetration is approached.

The computations were done using the heuristic described in Section 4.3.3.4. The results in Table 4.5 confirm the intuition that both smaller unit size and increasing DG penetration have positive effect on transportation losses.

The most remarkable feature of these results is the enormous sensitivity of the total losses to DG. Even for small penetration the losses drop to half the original value. Certainly for lower penetration, the size of the DG units makes little difference. Next it is notable that the first DG units are mostly placed in the inlands, but at a certain point Limburg takes over the central role.

The same computations were done for stochastic DG units: a placement is found for which expected transportation losses are minimized. For each case, the expected output

---

[15]Note that this is purely theoretical because of the following. In our model extra DG power is compensated by decreasing supply of the original supply nodes. This choice was made in order to approximate the situation in which all commercial power plants in the country take equal share in the compensation. When the share of DG becomes too large, this approximation fails.

4.3 Optimal placing of wind turbines

#DG units	30	60	120	6	12	24	3	6
unit size (MW)	100	100	100	500	500	500	2000	2000
total DG capacity (MW)	3000	6000	12000	3000	6000	12000	6000	12000
Coast and Brabant	0	0	1	0	0	0	0	0
Zuiderzee	0	0	6	0	0	1	0	0
Inlands	29	41	44	5	8	9	3	2
Limburg	0	9	48	0	2	10	0	3
Groningen Friesland	1	10	18	1	2	4	0	1
Network losses (MW)	3.71	1.60	0.04	3.72	1.59	0.06	1.82	0.46

Table 4.5: Dispersion of deterministic DG over wind zones.

of the DG units equals the size of its deterministic equivalent in the table above. Three scenarios were used: one with zero DG output, one with its expected output, and one with output of twice the expectation. The results are presented in Table 4.6.

#DG units	30	60	120	6	12	24	3	6
mean unit output (MW)	100	100	100	500	500	500	2000	2000
mean DG capacity (MW)	3000	6000	12000	3000	6000	12000	6000	12000
Coast and Brabant	0	0	22	0	0	5	0	1
Zuiderzee	0	1	6	0	0	1	0	0
Inlands	24	27	29	5	5	6	2	1
Limburg	2	22	47	0	5	9	1	3
Groningen Friesland	4	10	16	1	2	3	0	1
Network losses (MW)	4.287	3.17	3.0	4.3	3.209	3.0	3.651	4.055

Table 4.6: Dispersion of stochastic DG over wind zones.

It immediately stands out that the strong positive effect of deterministic DG on the transportation losses, is partly neutralized when DG units are stochastic. Interestingly, all penetrations yield comparable resulting net losses. Furthermore, stochastic DG units seem to be more dispersed than their deterministic equivalents. The strongest dispersion effect is seen for high penetration. However, it should be observed that for high expected penetration, there is excess of DG power in the third scenario (where all units produce twice their mean output). This results in the situation that the original supply nodes become demand nodes, making the North Sea Coast an attractive zone to place the DG units. The results for this situation should be distrusted.

4.3.5.3 Wind turbine placement

In this section results of the solutions to the wind turbine placement problem are presented and discussed. The computations were done using the heuristic discussed in Section 4.3.3.4. Recall that the maximum output power of the turbine that was regarded equals 2.5MW. The network results presented in the former section make clear that this quantity is almost negligible compared to the total load. Therefore, the output power was multiplied by successively 40, 200 and 800 in order to obtain interesting results. That is, the 'DG units' of maximum unit output of $100MW$, $500MW$ and $2000MW$ correspond to groups of 40, 200 and 800 large wind turbines.

The 729 scenarios that were simulated in Section 4.3.4 were used. These wind power scenarios reflect mutual correlation between the zones, and the Rayleigh distributed wind speeds for each zone apart. The resulting mathematical program is a MIQP with 64155 constraints and 88240 variables, of which 29 integer. Aimms' CPLEX 12.4 solver is able to solve this within a few minutes on a standard PC. It can be observed that both unit size and number of units have a negative influence on the computation time. (A larger number of units increases the number of basic feasible solutions strongly. It is less evident why larger unit size requires longer computation time.)

Consider the wind turbine placement problem. Placement of 30 wind turbines with maximum output of 100MW yields the result that all turbines are placed in Groningen, the average total output being less than 600MW. This shows that the solution of the wind turbine placement problem may differ from what one might expect when just wind or network data are considered. Highest wind speeds are to be found at the North Sea Coast, but there are no demand nodes there. Second highest wind speeds occur in the 'Zuiderzee' region which does have a shortage of 922MW. But apparently lower production in Groningen can cause a greater decrease in power transportation than a somewhat higher production in the 'Zuiderzee' region would cause. However for the Inlands and Limburg (which have an even more attractive position on the network for DG placement) wind speed is just too low.

#wind turbines	60	120	240	12	24	48	3	6
max unit output (MW)	100	100	100	500	500	500	2000	2000
mean DG capacity (MW)	1120	2060	3710	1140	2072	3750	1140	2140
Coast and Brabant	0	0	0	0	0	0	0	0
Zuiderzee	0	0	7	0	0	2	0	0
Inlands	0	1	57	0	0	11	0	0
Limburg	23	70	120	4	14	24	1	3
Groningen Friesland	37	49	56	8	10	11	2	3
Network losses (MW)	6.309	5.438	4.282	6.829	5.43	4.272	6.829	5.448
Solution time	157s	-	341s	-	-	-	374s	524s

Table 4.7: Dispersion of wind turbines over wind zones.

Table 4.7 below shows some results for higher penetrations. In the lowest row some computation times are displayed, which it took Aimms 3.12 on a computer with 4 GB RAM, Intel Duo core 3.0GHz processor, running under 64-bit Windows 7. It is interesting to see how the Inlands, which seemed so attractive both in the mean variance model for wind power, and in the network structure analysis, are less attractive in the wind turbine placement problem. Units are placed in the Inlands only after Groningen has reached a kind of saturation, and there are already many wind turbines in Limburg. This phenomenon might occur because the expected wind generation is simply too low in the Inlands; but seeing that the difference in expected outputs between Inlands and Limburg is smaller than between Groningen and Zuiderzee, some other reason seems likely.

When there are more than 30 units of 100MW (that is, 3000MW of installed wind capacity) in Groningen, there is excess of wind power in that region during the strong wind scenarios. Some of that power will flow to the neighbouring Inlands. Consequently, Limburg is more attractive than the Inlands because of its greater distance to Groningen.

Also neigbouring zones do not have full correlations: highest wind generation scenarios of one zone do not necessarily coincide with highest generation in the other zone. The probability of local power excess may therefore be decreased by placing units in the neigbouring zone. This may explain why the results for wind turbine placement show a higher spreading of generation units than the stochastic DG placement regarded in the former section (where correlations were 100%).

4.4 Summary

In this chapter two tactical planning problems in electricity networks have been considered. The first question was: what is an optimal mix of DGs so that energy loss is minimized? A Mixed Integer Quadratic Programming formulation (MIQP) was formulated that had to be simplified to solve. Depending on the various assumptions, such as whether it is allowed to transport overproduction to other districts or use the storage system, and whether there is additional demand from electric vehicles and heat pumps, we obtained different results from the model. This means that we have obtained several solutions in the case study each under these different assumptions. However, all the results indicate that implementing an optimal mix of DGs in the district can reduce energy loss substantially.

Another question was whether the use of DGs, heat pumps and electric vehicles creates capacity problems in the grid. If all houses have both types of DGs then problems will arise where the transformer and cables are overloaded. However, using our optimal solutions we did not come across any problems with overload. For the use of heat pumps and electric vehicles it is, unfortunately, more problematic. As long as heat pumps are distributed such that only one is connected to each phase overload does not occur. Most problems with overload come from the use of electric vehicles, which demand so much electricity that not even one electric vehicle can be charged at home without overloading the cables[16]. Hence, if one wants to include electric vehicles and heat pumps in a district, the cables and transformers must be reinforced to be able to handle such a large increase in demand.

The disadvantage of having both micro-CHP systems and PV solar panels as DGs, is that they generate electricity during the same period of the day. So micro-CHP systems and PV solar panels do not complement each other. However, each type of DG has a positive characteristic that can be used in specific situations. Because PV solar panels generate a relatively small amount of electricity, they can be used as a supplement to other generators. And, because micro-CHP systems generate a lot of electricity, they are useful when there is a large increase in demand, e.g. due to heat pumps and electric vehicles.

The results show that the higher the efficiency of a storage system the more it

---

[16]We do not mean the current type of electric vehicles here.

will be used to store overproduced electricity. Also, the larger the district the more efficient it becomes to use the storage system. If we let the storage system store as much overproduced electricity as possible without any control mechanism, the storage system will be charged and discharged every day. The consequences of this constant charge and discharge may be harmful to the storage system. Hence, when including a storage system in the district one needs to consider whether a control mechanism can be implemented. To see how this would affect the optimal mix one can include a control mechanism in the optimization model.

To be able to construct and solve the model a few assumptions had to be made on the data and on the estimation of losses. So there are some limitations on the model and results. This is especially true for the interpretation of losses. That is why we only use the estimated losses as an indication to check whether there is an improvement. In addition, the quality of electricity is not considered, which means that incorporating our mix of DGs may cause problems with power quality.

Using the optimal solutions with different input configurations we found that in all cases there is a big improvement compared to the grid without DGs. So our solutions can be used as a guideline for incorporating DGs in a district. These results also show that instead of arbitrarily deploying DGs in the district, it may be better to promote collaboration between home owners to invest in DGs so that large reductions in energy loss can be achieved. In this way it will be more profitable for the whole district.

We see a lot of possibilities for further research. First a solution method should be found that overcome the problems as discussed in Section 4.2.3. Next to this, we could introduce uncertainty in the model. One may be uncertain about the 'true' values of the coefficients in the model and can model these coefficients as random variables. In practice demand and production of electricity is non-deterministic. Even though they follow some predictable trend, there are some random elements. Hence, stochasticity can be incorporated into the model such that consumption and production follow some random distribution. One other aspect we left out is the so-called vehicle-to-grid. The idea behind a vehicle-to-grid system is that electric vehicles and other plug-in hybrid vehicles are used to store and release electricity when needed. Due to the fact that the storage systems of these vehicles have to be managed and controlled, modelling vehicle-to-grid is a part of the smart grid concept.

In the second problem in this chapter a normal load situation of a High Voltage (HV) network is studies and the optimal locations to build a given number of new wind mills has to be found, in order to minimize the expected energy transportation losses. Here we had to manage two trade-offs: local or central placement and spreading or concentration. A mathematical model was formulated in order to capture the wind power Generation Expansion Planning problem for loss minimization in a so-called two stage stochastic program. Network flow duality was applied such as to arrive at a NLP formulation for the stochastic Generation Expansion Planning problem for loss minimization. Rigorous solution algorithms for this NLP problem were outside the scope of this thesis. Therefore an simple heuristic was presented, together with a procedure to simulate appropriate scenarios for steady state wind power studies in multiple areas. This implementation of the model was used to generate results for the Dutch HV network.

From the computational results it was possible to conclude that the effect of reducing transmission losses by placing DG units is significantly weaker for stochastic than for deterministic units. Next to this, in order to achieve high transmission efficiency, stochastic DG units should be placed further apart than deterministic DG units, but this interesting effect occur only for higher penetration. From the specific Dutch situation we learned that the optimal distribution of wind energy over the Netherlands shows a higher spreading than can be explained merely from its stochasticity. The location dependent nature of wind energy is relevant for its effect on power grid performance. Because of the load situation of the Dutch HV grid Groningen and Friesland are the only regions with both attractive wind speeds, and where placement of wind turbines would result in reduction of transmission losses in the HV grid. However at the current scale of wind energy, a spreading strategy for wind turbines does not seem to be relevant, neither from a commercial point of view (variance reduction of total output for the Dutch market) nor from a technical point of view (reducing transmission losses)[17].

Apart from the computational results presented in this chapter, some general observations can be made:

- The loss minimizing uses of DG seem more relevant in large countries like US or Australia, than in a smaller and densely populated country like the Netherlands.

- Location selection of new wind turbines is done by individual commercial power companies, whereas costs for transportation are paid by all parties equally. Therefore in practice there is no financial incentive which stimulates locations that improve grid efficiency.

- LMP methods may open the possibility to let the causer of transportation losses pay for the loss. In such a scenario using DG for reducing losses could also have economic value.

---

[17]This concerns the HV grid: whether spreading strategies on a MV network scale would be desirable from a technical point of view, should be investigated in a different case study.

5 Summary

Electricity and telecommunications network providers operate in a turbulent period. The market is open for competition and the demand of their customers is changing dramatically, causing huge challenges for the network providers.

Information is reaching us more and more digitally. Mobile and fixed networks have to deal with this explosive growth of data traffic. The end of this is not yet in sight, mainly due to the increase in video services at higher resolution. For fixed networks this means that the default internet connection via ADSL (up to 8 Mb per second) is no longer sufficient. Providers using coaxial cable brought the fibre connection already far into the direction of their customers, so they can already offer 100 Mb per second, but this will soon no longer be sufficient. The incumbent telecommunications networks using twisted pair copper cables are not there yet. Mostly they are working on the rollout of VDSL, where the fibre connection is brought to the cabinet (Fibre to the Cabinet, FttCab) and also a speed of about 100 Mb can be achieved. That may be sufficient for now but this is also not future proof. In addition, they offer Fibre to the Home (FTTH) in certain areas. The latter roll out is not fast enough. In the past five years about 25% [138] of the Netherlands is reached, which was likely to be the easiest part to connect. Such a FttH connection costs around 1,000 euros which means a huge investment for the operators. This sketches the huge dilemma for the operators: bring a solution quickly that is not future proof or bring a future proof solution slowly and lose a part of your customers to the competition. In this thesis, we provide a systematic solution for this dilemma.

Electricity networks have to handle another dilemma. They face an increasing demand due to, e.g., electric cars, and they see their customers (households) partly generate their own electricity. They are called *prosumers*, a combination of producer and consumer. Solar panels are common, but also more and more electricity generating boilers (known as micro-CHP) and small wind turbines are placed in residential areas. This production is also highly variable, while they are depending on sun and wind. These trends lead for the network managers to the challenge how to keep their network in balance. Create your own flexible generation, using (price) incentives to influence the demand of consumers and smart grids are all investigated and you always have the very expensive solution to strengthen and renew the network cables. In this thesis we looked for a solution that is at the tactical level. How many of each type of local generator do you need and where do you place them to minimize transmission losses and the losses by demand-supply mismatch?

For both problem areas we studied (tactical) planning issues, all of which are very complex to calculate. The operational application of our algorithms also determine the scope of the presented solution in this thesis. The focus is on efficient yet fairly accurate methods rather than methods that generate an optimal solution in long calculation times.

5.1 Telecom infrastructure

The above mentioned dilemma for telecom operators requires a deployment strategy that satisfies customer demand as much as possible, but is also economically feasible. Responding to the customers needs all the time may be too expensive, however, a cheap deployment neglecting these needs can cause that there are no customers left at a certain moment, as all of them have switched to a competitor that can offer the demanded bandwidth. In this thesis we presented an economic model that can calculate the effect of all kinds of strategies. Analyses led to the conclusion that a migration of Full Copper via FttCab (VDSL-based) and FttCurb (G.Fast-based) to FTTH is both cost effective and retains the customers.

For this migration, it is important to build up the network configurations considering the next one. When planning the current roll-out, you should take into account as much as possible the successive infrastructures. We gave an example of this in this thesis. Then we presented for each of the migration steps (Full Copper to FttCab, FttCab to FttCurb and FttCurb/Full Copper to FttH) a planning methodology.

For the first migration step, FttCab planning, we presented a planning methodology, primarily based on three steps. First was determined which street cabinets should be provided with active equipment. The aim is to chose a minimum number of cabinets from a cost point of view, which cover, e.g., 95% of the connected houses. Then the activated cabinets must be connected via a fibre optic ring. To this end, in step 2, first clusters of activated cabinets are created of a certain maximum size, respecting the capacity of the fibre optic ring. Next, in step 3, the route of these rings is determined. All three problems are already difficult to solve (NP-hard). For all three problems, we proposed a heuristic approach and showed how well they work. It is important here that we are not so much interested in the best solution, but in a good solution very quickly, preferably within seconds, minutes at most. For step 1 a greedy heuristic was developed that gave a good solution for all test cases within one second. For step 2 we developed a variant of Lloyd's algorithm for solving k-means clusters, taking seconds to solve bigger problems. In step 3 it is important to create fibre rings that do not use the same piece of track twice. For this, we developed a heuristic that is based on Dijkstra's algorithm and a Insertion Algorithm for solving Travelling Salesman Problems which creates an initial solution quickly and then solves the double used tracks in a greedy way, also within seconds.

To see how well the three-step approach works, we then created a method that solves the second and third step simultaneously. As expected, this method provides better solutions, since the routing information is included in the clustering. Also as expected, the computation time of this method was longer, but less longer than expected.

For the second migration step, FttCurb planning, we identified three main choices that you have to make if you want to create such a network, based on the technology G.Fast. First, the network operator must decide whether or not he wants to connect all the houses within a certain distance, for G.Fast the desired distance is 200 meters, or a certain percentage of the houses. Then he has to determine whether the new active nodes have a capacity limitation, or whether this location can be expanded unlimited. Finally, he must decide whether you want to connect the new active nodes via a fibre optic star or tree or via a ring structure. If every choice has two options, then there are a total of eight possible configurations of the network. In this thesis we described for each of the eight options the planning methodology. Then we elaborated the planning for two Dutch cities, Amsterdam and The Hague, to estimate the cost of the network. The results of this study have also acted again as validation of the aforementioned (simplified) economic model.

Finally, the third step involves the migration to FttH. Here we showed how the location of the new central node (PoP) can be determined. We extended a simple heuristic, called JMS heuristic, on a number of points to bring more details in this method. We then considered the real path of the fibres to the homes starting at this PoP locations. This seems like a trivial question, but you can get much profit here. We described a new method to determine these paths and developed an addition to incorporate 'smart co-laying'. In the method the social costs of inconvenience, turn-over loss etc. are taken into account and can be minimized together with the investment costs. The 'smart co-laying' options give the possibility of a phased construction to use other infrastructural project as much as possible to lay the fibre in the ground together with other planned work as street maintenance and sewer renewals, all minimizing the social costs.

5.2 Electricity infrastructure

The producers of electricity and the managers of the networks have to deal with increasing demand and an increase in highly variable production by local generators such as solar cells and wind turbines. In this thesis we described two studies that study the tactical planning of these generators. How many of which kind and where do you have to place them to minimize losses. Local generation can reduce transmission losses because it does not have to be transported far. On the other hand, this local generation introduces large variability in supply, which also need to be transported again, needs to be stored (not efficiently) or will be lost through lack of demand. When the wind turbines are spread, they have lower variability due to a lower correlation between the wind at different places. If the wind blows in Middelburg, then it will also blow in Vlissingen. That it blows that hard in Groningen is a lot less certain.

The first question that we had to answer here was: What is an optimal mix of DGs in a district so that energy loss is minimized? We developed a Mixed Integer Quadratic Programming formulation (MIQP) that we had to simplify to solve it. We obtained several solutions in the case studies each under different assumptions. However, the results indicate that implementing an optimal mix of DGs in the district can reduce energy loss substantially. Next to this we could show that using our optimal solutions will not cause problems with overload. For the use of heat pumps and electric vehicles it is, unfortunately, more problematic. Most problems with overload come from the use of electric vehicles, which demand so much electricity that not even one electric vehicle can be charged at home without overloading the cables. Finally we saw that however both micro-CHP systems and PV solar panels generate electricity during the same period of the day, each type of DG has a positive characteristic that can be used in specific situations. Because PV solar panels generate a relatively small amount of electricity, they can be used as a supplement to other generators. And, because micro-CHP systems generate a lot of electricity, they are useful when there is a large increase in demand for example due to heat pumps and electric vehicles. Using the optimal solutions with different input configurations we found that in all cases there is a big improvement compared to the grid without DGs. So our solutions can be used as a guideline for incorporating DGs in a district. These results also show that instead of arbitrarily deploying DGs in the district, it may be better to promote collaboration between home owners to invest in DGs such that large reductions in energy loss can be achieved. In this way it will be more profitable for the whole district.

In the second problem we studied a normal load situation of a High Voltage (HV) network and tried to find the optimal locations to build a given number of new wind turbines, in order to minimize the expected energy transportation losses. Here we had to manage two trade-offs: (1) local or central placement and (2) spreading or concentration. A mathematical model was formulated in order to capture the wind power Generation Expansion Planning problem for loss minimization in a so-called two stage stochastic program. Network flow duality was applied such as to arrive at a NLP formulation for the stochastic Generation Expansion Planning problem for loss minimization. Rigorous solution algorithms for this NLP problem were outside the scope of this thesis. Therefore a simple heuristic was presented, together with a procedure to simulate appropriate scenarios for steady state wind power studies in multiple areas. This implementation of the model was used to generate results for the Dutch HV network. From the computational results it was possible to conclude that the effect of reducing transmission losses by placing DG units is significantly weaker for stochastic than for deterministic units. Next to this, in order to achieve high transmission efficiency, stochastic DG units should be placed further apart than deterministic DG units, but this interesting effect occur only for higher penetration. From the specific Dutch situation we learned that the optimal distribution of wind energy over the Netherlands shows a higher spreading than can be explained merely from its stochasticity. The location dependent nature of wind energy is relevant for its effect on power grid performance. Because of the load situation of the Dutch HV grid Groningen and Friesland are the only regions with both attractive wind speeds, and where placement of wind turbines would result in reduction of transmission losses in the HV grid. However at the current scale of wind energy, a spreading strategy for wind turbines does not seem to be relevant, neither from a commercial point of view (variance reduction of total output for the Dutch market) nor from a technical point of view (reducing transmission losses).

Bibliography

[1] C. Ababei and R. Kavasseri. Efficient network reconfiguration using minimum cost maximum flow-based branch exchanges and random walks-based loss estimations. *IEEE Transactions on Power Systems*, 326(1), 2011.

[2] K. Ahmed, P. Verhagen, F. Phillipson, and C. Smit. *4GBB Techno-Economic Feasibility and Regulatory Issues, Part 3 - Techno Economic Modeling - TNO approach*. TNO - CELTIC/4GBB project, 2012.

[3] R.K. Ahuja, T.L. Magnanti, and J.B. Orlin. *Network Flows*. Prentice-Hall, Englewood Cliffs, 1993.

[4] J. Alleman. A new view of telecommunications economics. *Telecommunications Policy*, 26:87–92, 2002.

[5] D. Aloise, A. Deshpande, P. Hansen, and P. Popat. Np-hardness of Euclidean sum-of-squares clustering. *Machine Learning*, 75:245–249, 2009.

[6] M. Andrews and L. Zhang. Hardness of the undirected edge-disjoint paths problem. In *Proceedings of the thirty-seventh annual ACM symposium on Theory of computing*, pages 276–283, 2005.

[7] C.H. Antunes, J.F. Craveirinha, and J.N. Clímaco. Planning the evolution to broadband access networks: A multicriteria approach. *European Journal of Operational Research*, 109(2):530–540, 1998.

[8] D. Arthur and S. Vassilvitskii. k-means++: The advantages of careful seeding. pages 1027–1035, 2007.

[9] V. Arya, N. Garg, R. Khandekar, A. Meyerson, K. Munagala, and V. Pandit. Local search heuristics for k-median and facility location problems. In *Proceedings of the 33rd ACM Symposium on Theory of Computing*, 2001.

[10] R. Baldacci, M. Dell'Amico, and J. Salazar González. The capacitated m-ring-star problem. *Operations Research*, 55(6):1147–1162, 2007.

[11] M.E. Baran and F.F. Wu. Network reconfiguration in distribution systems for loss reduction and load balancing. *IEEE Transactions on Power Delivery*, 4(2): 1401–1407, 1989.

[12] L. S. Belyaev. Pay-off matrix technique. *Energy*, 15(7/8):631–643, 1990.

[13] D. Bertsimas and J.N. Tsitsiklis. *Introduction to Linear Optimization*. Athena Scientific, Belmont, Massachusetts, 1997.

[14] R. Bhandari. *Survivable Networks: Algorithms for Diverse Routing*. Kluwer Academic Publishers, 1999.

[15] A. Bley, I. Ljubić, and O. Maurer. Lagrangian decompositions for the two-level fttx network design problem. *EURO Journal on Computational Optimization*, 1(3-4):221–252, 2013.

[16] P. Bradley, K.P. Bennet, and A. Demiriz. Constrained k-means clustering. Technical report, Microsoft Research MSR-TR-2000-65, 2000.

[17] R. Brennenraedts, B. Minne, R. te Velde, and J Veldkamp. De bijdrage van de telecomsector aan de economische groei in nederland (in dutch). Technical report, Dialogic, 2012.

[18] R.F.M. van den Brink. Enabling 4GBB via the last copper drop of a hybrid FttH deployment. Technical report, TNO, The Netherlands, 2011.

[19] R.F.M. van den Brink. Enabling 4gbb via the last copper drop of a hybrid FttH deployment. *Broadband Journal of the SCTE*, 33(2):40–46, 2011.

[20] R.F.M. van den Brink and F. Phillipson. *4GBB Techno-Economic Feasibility and Regulatory Issues, Part 4 - Techno Economic Calculations - TNO approach*. TNO - CELTIC/4GBB project, 2012.

[21] G. Carpinelli, G. Celli, F. Pilo, and A. Russo. Distributed generation sizing and siting under uncertainty. In *Proceedings of Power Tech*, volume 4, pages 376–382. IEEE, 2001.

[22] K. Casier. *Techno-Economic Evaluation of a Next Generation Access Network Deployment in a Competitive Setting*. Universiteit Gent, 2009.

[23] CELTIC-4GBB. URL www.4gbb.eu.

[24] CELTIC-ECOSYS. URL www.celtic-initiative.org/projects/ecosys.

[25] M. Chardy, M-C. Costa, A. Faye, and M. Trampont. Optimizing splitter and fiber location in a multilevel optical FttH network. *European Journal of Operational Research*, 222(3):430–440, 2012.

[26] R. Chedid, T. Mezher, and C. Jarrouche. A fuzzy programming approach to energy resource allocation. *International Journal of Energy Research*, 23:303–317, 1999.

[27] R. Li Yang Chen. *Models and Algorithms for Stochastic Network Design and Flow Problems: Applications in Truckload Procurement Auctions and Renewable Energy*. PhD Thesis, University of Michigan, 2010.

[28] G. Clarke and J.W. Wright. Scheduling of vehicles from a central depot to a number of delivery points. *Operations Research*, 12(4):pp. 568–581, 1964.

[29] L. Cooper. Heuristic methods for location-allocation problems. *SIAM Review*, 6: 37–53, 1964.

[30] T.H. Cormen, C.E. Leiserson, R.L. Rivest, and C. Stein. *Introduction to Algorithms*. The MIT Press, Cambridge, Massachusetts, USA, 3rd edition, 2009.

[31] D. Crevier. *Approximate Transmission Network Models for Use in Analysis and Design*. MIT Energy Laboratory Report, 1972.

[32] G.A. Croes. A method for solving traveling salesman problems. *Operations Research*, 6:791–812, 1958.

[33] N.N. Croes. *Impact of distributed energy generation on energy loss: finding the optimal mix*. MSc Thesis, University of Groningen and TNO, 2011.

[34] N.N. Croes and F. Phillipson. Tactical planning of distributed generation: Finding the optimal mix to reduce energy loss. *Submitted*, 2013.

[35] N.N. Croes, F. Phillipson, and M.A. Schreuder. Tactical congestion management: The optimal mix of decentralised generators in a district. In *Proceedings of Integration of Renewables into the Distribution Grid, CIRED 2012 Workshop*, 2012.

[36] R. Daamen and F. Phillipson. Comparison of heuristic methods for the design of edge disjoint circuits. *Submitted*, 2013.

[37] K. Darby-Dowman and H.S. Lewis. Lagrangian relaxation and the single-source capacitated facility-location problem. *The Journal of the Operational Research Society*, 39(11):1035–1040, 1988.

[38] A.R. De Musgrove. A linear programming analysis of liquid-furl production and use options for Australia. *Energy*, 9:281–302, 1984.

[39] E.D. Demaine, M. Hajiaghayi, and P.N. Klein. Node-weighted Steiner tree and group Steiner tree in planar graphs. In *Automata, Languages and Programming*, volume 5555 of *Lecture Notes in Computer Science*, pages 328–340. Springer Berlin Heidelberg, 2009.

[40] E.W. Dijkstra. A note on two problems in connexion with graphs. *Numerische Mathematik*, 1(1):269–271, 1959.

[41] EIA. URL http://www.eia.gov/electricity/state/.

[42] A. Eira, J. Pedro, and J. Pires. Optimized design of multistage passive optical networks. *Journal of Optical Communications and Networking*, 4(5):402–411, 2012.

[43] EnergieNed. *Electricity Distribution Grids*. Kluwer Techniek Elektro/Elektronica, 1996. (in Dutch).

[44] M. Esmaili, E. Firozjaee, and H. Shayanfar. Optimal placement of distributed generations considering voltage stability and power losses with observing voltage-related constraints. *Applied Energy*, 113:1252–1260, 2014.

[45] A. Faruqui, S. Sergici, and L. Akaba. Dynamic pricing of electricity for residential customers: the evidence from Michigan. *Energy Efficiency*, 6:571–584, 2013.

[46] A.E. Feijoo, J. Cidras, and J.L.G. Dornelas. Wind speed simulation in wind farms for steady-state security assessment of electrical power systems. *IEEE Transactions on Energy Conversions*, 14(4):1582–1588, 1999.

[47] A. Fink, G. Schneidereit, and S. Voß. Solving general ring network design problems by meta-heuristics. In *Computing Tools for Modeling, Optimization and Simulation*, pages 91–113. Springer, 2000.

[48] S. Fortune, J. Hopcroft, and J. Wyllie. The directed subgraph homeomorphism problem. *Theoretical Computer Science*, 10(2):111–121, 1980.

[49] M.R. Garey, R.L. Graham, and D.S. Johnson. The complexity of computing Steiner minimal trees. *SIAM Journal on Applied Mathematics*, 32(4):835–859, 1977.

[50] M. Gendreau, M. Labbé, and G. Laporte. Efficient heuristics for the design of ring networks. *Telecommunication Systems*, 4(1):177–188, 1995.

[51] M. Gitizadeh, A.A. Vahed, and J. Aghaei. Multistage distribution system expansion planning considering distributed generation using hybrid evolutionary algorithms. *Applied Energy*, 101:655–666, 2012.

[52] I. Gódor and G. Magyar. Cost-optimal topology planning of hierarchical access networks. *Computers & operations research*, 32(1):59–86, 2005.

[53] M.X. Goemans and D.P. Williamson. A general approximation technique for constrained forest problems. *SIAM Journal on Computing*, 24(2):296–317, 1995.

[54] S. Gollowitzer and I. Ljubić. Mip models for connected facility location: A theoretical and computational study. *Computers and Operations Research*, 38:435–449, 2011.

[55] S. Gollowitzer, Gouveia L., and I. Ljubić. A node splitting technique for two level network design problems with transition nodes. *Lecture Notes in Computer Science*, 6701:57–70, 2011.

[56] S. Gollowitzer, B. Gendron, and I. Ljubić. Capacitated network design with facility location. Technical report, University of Vienna, 2012.

[57] Y. Gong, C. Gan, C. Wu, and Ru. Wang. Novel ring-based wdm-pon architecture with high-reliable remote nodes. *Telecommunication Systems*, 2013.

[58] H. J. Greenberg. Greedy algorithms for minimum spanning tree. Technical report, University of Colorado at Denver - http://glossary.computing. society.informs.org/notes/spanningtree.pdf, 1998.

[59] I. Hadjipaschalis, A. Poullikkas, and V. Efthimiou. Overview of current and future energy storage technologies for electric power applications. *Renewable and Sustainable Energy Reviews*, 13(6-7):1513–1522, 2009.

[60] M. Haouari, S. Bhar Layeb, and H.D. Sherali. Algorithmic expedients for the prize collecting Steiner tree problem. *Discrete Optimization*, 7:32–47, 2010.

[61] Å. L. Hauge, J. Thomsen, and E. Löfström. How to get residents/owners in housing cooperatives to agree on sustainable renovation. *Energy Efficiency*, 6: 315–328, 2013.

[62] M. Henningsson, K. Holmberg, M. Rönnqvist, and P. Värbrand. Ring network design by lagrangian based column generation. *Telecommunication Systems*, 21 (2-4):301–318, 2002.

[63] R.B. Hiremath, S. Shikha, and N.H. Ravindranath. Decentralized energy planning; modeling and application - a review. *Renewable and Sustainable Energy Reviews*, 11:729–752, 2007.

[64] K. Holmberg, M. Ronnqvist, and D. Yuan. An exact algorithm for the capacitated facility location problems with single sourcing. *European Journal of Operational Research*, 113:544–559, 1999.

[65] K. Hornik and B. Grün. TSP-infrastructure for the traveling salesperson problem. *Journal of Statistical Software*, 23(2):1–21, 2007.

[66] Z. Hu and W.T. Jewell. Optimal generation expansion planning with integration of variable renewables and bulk energy storage systems. In *Technologies for Sustainability (SusTech), 2013 1st IEEE Conference on*, pages 1–8. IEEE, 2013.

[67] M. Inaba, N. Katoh, and H. Imai. Applications of weighted Voronoi diagrams and randomization to variance-based k-clustering. In *Proceedings of 10th ACM Symposium on Computational Geometry*, pages 332–339, 1994.

[68] IST-TONIC. URL www.ist-tonic.org.

[69] E. Ivanova, N.I. Voropai, and E. Handschin. A multi-criteria approach to expansion planning of wind power plants in electric power systems. In *Proceedings of Power Tech, 2005 IEEE Russia*. IEEE, 2005.

[70] A.K. Jain. Data clustering: 50 years beyond k-means. *Pattern Recognition Letters*, 31:651–666, 2010.

[71] S. Jebaraj and S. Iniyan. A review of energy models. *Renewable and Sustainable Energy Reviews*, 10(4):281–311, August 2006.

[72] D.R. Jensen. A generalization of the multivariate rayleigh distribution. *The Indian Journal of Statistics*, A(32):193–208, 1970.

[73] D.R. Jensen. Multivariate distributions having weibull properties. *Journal of Multivariate Analysis*, 9:267–277, 1979.

[74] R.A. Johnson and D.W. Wichern. *Applied Multivariate Statistical Analysis*. Prentice-Hall, Upper Saddle River, 2007.

[75] M.T. Kalsch, M.F. Koerkel, and R. Nitsch. Embedding ring structures in large fiber networks. In *Proceedings of Telecommunications Network Strategy and Planning Symposium (NETWORKS)*, 2012.

[76] R.M. Karp. *Reducibility Among Combinatorial Problems*. Springer, 1972.

[77] G.A.P. Kindervater and M.W.P. Savelsbergh. Vehicle routing: Handling edge exchanges. In E. Aarts and J. Lenstra, editors, *Local Search in Combinatorial Optimization*, section 10, pages 337–360. Wiley, 1997.

[78] P. Klein and R. Ravi. A nearly best-possible approximation algorithm for node-weighted Steiner trees. *Journal of Algorithms*, 19:104–115, 1995.

[79] Y. Kobayashi and C. Sommer. On shortest disjoint paths in planar graphs. *Discrete Optimization*, 7(4):234 – 245, 2010. ISSN 1572-5286.

[80] J.K. Kok, C.J. Warmer, and I.G. Kamphuis. Powermatcher: multiagent control in the electricity infrastructure. In *Proceedings of the fourth international joint conference on Autonomous agents and multiagent systems*, pages 75–82. ACM, 2005.

[81] K. Kok. *The PowerMatcher: Smart Coordination for the Smart Electricity Grid*. PhD Thesis, Vrije Universiteit Amsterdam and TNO, 2013.

[82] A. Kokangul and A. Ari. Optimization of passive optical network planning. *Applied Mathematical Modelling*, 35(7):3345–3354, 2011.

[83] L. Kools. *The effect of modeling choices on decision making regarding the optimal planning of distributed generation*. MSc Thesis, University of Groningen and TNO, 2014.

[84] G.M. Kopanos, M.C. Georgiadis, and E.N. Pistikopoulos. Energy production planning of a network of micro combined heat and power generators. *Applied Energy*, 102:1522–1534, 2012.

[85] B. Korte, L. Lovasz, H. Jurgen Promel, and A. Schrijver. *Paths, Flows, and VLSI-Layout*. Springer-Verlag New York, Inc., Secaucus, NJ, USA, 1990.

[86] L. Kou, G. Markowsky, and L. Berman. A fast algorithm for Steiner trees. *Acta Informatica*, 14:141–145, 1981.

[87] C. Kraemer, K. Goldermann, C. Breuer, P. Awater, and A. Moser. Optimal positioning of renewable energy units. In *Proceedings of Energytech*. IEEE, 2013.

[88] M. Labbé, G. Laporte, I. Rodríguez Martín, J. José, and J.S. González. The ring star problem: polyhedral analysis and exact algorithm. *Networks*, 43:177–189, 2004.

[89] W.C. Labys and G. Kuczmowski, T.and Infanger. Special programming models. *Energy*, 15(7/8):607–617, 1990.

[90] G. Laporte, M. Gendreau, J-Y Potvin, and F. Semet. Classical and modern heuristics for the vehicle routing problem. *International Transactions in Operational Research*, 7(4-5):285–300, 2000.

[91] R.C. Larson and A.R. Odoni. *Urban Operations Research*. Prentice Hall, 1981.

[92] A.D.T. Le, M.A. Kashem, M. Negnevitsky, and G. Ledwich. *Optimal Distributed Generation parameters for reducing losses with economic consideration*. Research paper for the Australian Research Council, 2007.

[93] T. Leenman and F. Phillipson. Optimal placing of wind turbines: modelling the uncertainty. *To appear in: Journal of Clean Energy Technologies*, 2014.

[94] J.K. Lenstra and A.H.G. Rinnooy Kan. Complexity of vehicle routing and scheduling problems. *Networks*, 11(2):221–227, 1981.

[95] J. Li and G. Shen. Cost minimization planning for passive optical networks. In *Proceedings of Conference on Optical Fiber communication/National Fiber Optic Engineers Conference*, 2008.

[96] L. Lima Pinto and G. Laporte. An efficient algorithm for the Steiner tree problem with revenu, bottleneck and hop objective function. *European Journal of Operational Research*, 207:45–49, 2010.

[97] S.P. Lloyd. Least squares quantization in pcm. *IEEE Transactions on Information Theory*, IT-28(2):129–137, 1982.

[98] M.L. Luhanga, M.J. Mwandosya, and P.R. Lutegenya. Optimisation in computerized energy modeling for Tanzania. *Energy*, 18:1171–1179, 1993.

[99] M. Lv and X. Chen. Heuristic based multi-hierarchy passive optical network planning. In *Proceedings of 5th International Conference on Wireless Communications, Networking and Mobile Computing*, 2009.

[100] E. Lysen, S. van Egmond, and S. Hagedoorn. Electricity storage: Status and prospects for the Netherlands. Technical report, SenterNovem, 2006. Available at www.uce-uu.nl (in Dutch).

[101] J. MacQueen. Some methods for classification and analysis of multivariate observations. In *Proceedings of 5th Berkeley Symposium on Mathematical Statistics and Probability*, pages 281–297, 1967.

[102] M. Mahdian, Y. Ye, and J. Zhang. Approximation algorithms for metric facility location problems. *SIAM Journal on Computing*, 36(2):411–432, 2003.

[103] E. Mashhour and S.M. Moghaddas-Tafreshi. Integration of distributed energy resources into low voltage grid: A market-based multiperiod optimization model. *Electric Power Systems Research*, 80(4):473–480, 2009.

[104] Analysys Mason. *Conceptual approach for the fixed and mobile BULRIC models.* OPTA, 2010.

[105] G.R. Mateus, F.R.B. Cruz, and H.P.L. Luna. An algorithm for hierarchical network design. *Location Science*, 2(3):149–164, 1994.

[106] K. Mehlhorn. A faster approximation algorithm for the Steiner problem in graphs. *Information Processing Letters*, 27:125–128, 1988.

[107] S. Melkote and M.S. Daskin. Capacitated facility location/network design problems. *European Journal of Operational Research*, 129:481–495, 2001.

[108] V.H. Méndez, J. Rivier, J.I. de la Fuente, T. Gómez, J. Arcéluz, and J. Marín. Impact of distributed generation on distribution losses. In *Proceedings of the 3rd Mediterranean Conference and Exhibition on Power Generation, Transmission, Distribution and Energy Conversion*, November 2002. Available at www.iit.upcomillas.es.

[109] M. Mitcsenkov, G. Paksy, and T. Cinkler. Topology design and capex estimation for passive optical networks. In *Proceedings of BROADNETS 2009*, 2009.

[110] T. Monath, N.K. Elnegaard, P. Cadro, D. Katsianis, and D. Varoutas. Economics of fixed broadband access network strategies. *IEEE Communications Magazine*, 41(9):132–139, 2003.

[111] Z. Naji-Azimi, M. Salari, and P. Toth. A heuristic procedure for the capacitated m-ring-star problem. *European Journal of Operational Research*, 207(3):1227–1234, 2010.

[112] George A Orfanos, Pavlos S Georgilakis, and Nikos D Hatziargyriou. Transmission expansion planning of systems with increasing wind power integration. 2012.

[113] A. Ouali, K.F. Poon, and A. Chu. FttH network design under power budget constraints. In *Proceedings of the Internation Symposium on Integrated Network Management*, pages 748–751. IEEE, 2013.

[114] G. Papaefthymiou and B. Kloeckl. Mcmc for wind power simulation. *IEEE Transactions on Energy Conversion*, 23(1):234–240, 2008.

[115] G. Pepermans, J. Driesen, D. Haeseldonckx, R. Belmans, and W. Díhaeseleer. Distributed generation: Definition, benefits and issues. *Energy Policy*, 33(6):787–798, 2005.

[116] A.L. Peressini, F.E. Sullivan, and J.J. Uhl. *The Mathematics of Nonlinear Programming*. Springer-Verlag New York Inc., 1993.

[117] F. Phillipson. Efficient clustering of cabinets at FttCab. In *Internet of Things, Smart Spaces, and Next Generation Networking*, pages 201–213. Springer Berlin Heidelberg, 2013.

[118] F. Phillipson. A cost effective topology migration path towards fibre. In *Proceedings of the 3rd International Conference on Information Communication and Management (ICICM2013)*, Paris, France, 2013.

[119] F. Phillipson. Planning of fibre to the curb using g.fast in multiple rollout scenarios. In *Proceedings of the 3rd International Conference on Information Communication and Management (ICICM2013)*, Paris, France, 2013.

[120] F. Phillipson. Connecting the houses at FttH with respect for social costs: solving a streiner tree problem with timing benefits. *Submitted*, 2013.

[121] F. Phillipson. Roll-out of reliable fibre to the cabinet: an interactive planning approach. *Submitted*, 2013.

[122] F. Phillipson. Fast roll-out of fibre-to-the-cabinet: optimal activation of cabinets. *Submitted*, 2014.

[123] F. Phillipson, C. Smit-Rietveld, and P. Verhagen. Fourth generation broadband delivered over copper - a techno-economic study. *Journal of Optical Communications and Networking*, 5:1328–1342, 2013.

[124] R.C. Prim. Shortest connection networks and some generalizations. *Bell Systems Technical Journal*, 36:1389–1401, 1957.

[125] K. Purchala, L. Meeus, D. Van Dommelen, and R. Belmans. Usefulness of dc power flow for active power flow analysis. In *Proceedings of IEEE Power Energy Society General Meeting*, 2005.

[126] V.H. Mendez Quezada, J.R. Abbad, and T.G. San Roman. Assessment of energy distribution losses for increasing penetration of distributed generation. *IEEE Transactions on Power Systems*, 21(2), 2006.

[127] R. Srinivasa Rao and S.V.L. Narasimham. Optimal capacitor placement in a radial distribution system using plant growth simulation algorithm. *World Academy of Science, Engineering and Technology*, 33(45):1133–1139, 2008.

[128] Rijksoverheid. URL http://www.rijksoverheid.nl/onderwerpen/duurzame-energie/meer-duurzame-energie-in-de-toekomst.

[129] T. Rokkas, D. Katsianis, and D. Varoutas. Techno-economic evaluation of FttC/VDSL and FttH roll-out scenarios: Discounted cash flows and real option valuation. *Journal of Optical Communications and Networking*, 2(9):760–771, 2010.

[130] R.T. Rust and A.J. Zahorik. Customer satisfaction, customer retention, and market share. *Journal of Retailing*, 69(2):193–215, 1993.

[131] P.S. Satsangi and E.A.S. Sarma. Integrated energy planning model for India with particular reference to renewable energy prospects. In *Energy options for the 90's: proceedings of the National Solar Energy Convention held at Indian Institute of Technology*, pages 596–620, New Delhi, 1988. Tata McGraw Hill.

[132] M.J.J. Scheepers and A.F. Wals. New approach in electricity network regulation. An issue on effective integration of distributed generation in electricity supply systems. Technical Report ECN-C–03-107, Energy Research Centre of the Netherlands (ECN), September 2003. Available at www.ecn.nl.

[133] S.M. Schoenung. Characteristics and technologies for long- vs. short-term energy storage. A study by the DOE energy storage systems program. A Study by the DOE Energy Storage Systems Program SAND2001-0765, Sandia National Laboratories, March 2001.

[134] D.B. Shmoys, E. Tardos, and K.I. Aardal. Approximation algorithms for facility location problems. In *Proceedings of the twenty-ninth annual ACM symposium on Theory of computing*, pages 265–274. ACM, 1997.

[135] D. Singh, D. Singh, and K.S. Verma. GA based energy loss minimization approach for optimal sizing and placement of distributed generation. *International Journal of Knowledge-Based and Intelligent Engineering Systems*, 12:147–156, 2008.

[136] M. Steinbach. Markowitz revisited: Mean-variance models in financial portfolio analysis. *SIAM Review*, 43:31–85, 2001.

[137] B. Stott, J. Jardim, and O. Alsac. Dc power flow revisited. *IEEE Transactions on Power Systems*, 24(3):1290–1300, 2009.

[138] Stratix. *FTTH monitor 2013/Q3*. Stratix Hilversum, 2013.

[139] L. Suganthi and T.R. Jagadeesan. A modified model for prediction of Indiaï£¡s future energy requirement. *International Journal of Energy and Environment*, 3: 371–386, 1992.

[140] J.W. Suurballe and R.E. Tarjan. A quick method for finding shortest pairs of disjoint paths. *Networks*, 14(2):325–336, 1984.

[141] W.S. Tan, M.Y. Hassan, M.S. Majid, and H. Abdul Rahman. Optimal distributed renewable generation planning: A review of different approaches. *Renewable and Sustainable Energy Reviews*, 18:626–645, 2013.

[142] A. Telkamp. Optimalisatie van de netwerkstructuur voor fiber to the home. Master's thesis, University of Groningen and TNO, January 2005.

[143] T. Thomadsen and T. Stidsen. Hierarchical ring network design using branch-and-price. *Telecommunication Systems*, 29(1):61–76, 2005.

[144] H.C. Tijms. *A First Course in Stochastic Models*. Wiley, 2003.

[145] D. Tipper. Resilient network design: challenges and future directions. *Telecommunication Systems*, 2013.

[146] TNO. *Vraag en aanbod Next-Generation Infrastructures 2010 tot 2020 (in Dutch)*. TNO (online available), 2010.

[147] P. Tseng and D. Bertsekas. An ϵ-relaxation method for separable convex cost generalized network flow problems. In *Proceedings of the 5th International IPCO conference, Vancouver*, 1996.

[148] K. Vajanapoom, D. Tipper, and S. Akavipat. Risk based resilient network design. *Telecommunication Systems*, 52:799–811, 2012.

[149] S. Verbrugge, K. Casier, B. Lannoo, C. Mas Machuca, T. Monath, M. Kind, and M. Forzati. Research approach towards the profitability of future FttH business models. In *Proceedings Telecommunications Network Strategy and Planning Symposium (NETWORKS)*, Warshaw, Poland, June 2011.

[150] Vergilius. Aeneis. 5:231.

[151] C. Wang and M. Hashem Nehrir. Analytical approaches for optimal placement of distributed generation sources in power systems. *IEEE Transactions on Power Systems*, 19(4), 2004.

[152] B.M. Waxman. Routing of multipoint connections. *IEEE Journal on Selected Areas in Communications*, 6(9):1617–1622, 1988.

[153] J. Xu, S.Y. Chiu, and F. Glover. Optimizing a ring-based private line telecommunication network using tabu search. *Management Science*, 45(3):330–345, 1999.

[154] B. Yang and S.Q. Zheng. Finding min-sum disjoint shortest paths from a single source to all pairs of destinations. In *Proceedings of the Third international conference on Theory and Applications of Models of Computation*, TAMC'06, pages 206–216, Berlin, Heidelberg, 2006. Springer-Verlag.

[155] R. Zhao, H. Liu, and R. Lehnert. Topology design of hierarchical hybrid fiber-vdsl access networks with ACO. In *Proceedings of Fourth Advanced International Conference on Telecommunications*, 2008.

[156] R. Zhao, L. Zhou, and C. Mas Machuca. Dynamic migration planning towards FttH. In *Telecommunications Network Strategy and Planning Symposium (NETWORKS)*, pages 1617–1622, Warshaw, Poland, September 2010.

Nederlandse samenvatting

Efficiënte algoritmes voor infrastructurele netwerken: planningsproblemen en economische impact

Leveranciers van elektriciteit en telecommunicatiediensten zitten in een roerige periode. De markt ligt open voor concurrentie en de vraag van klanten verandert enorm. Dit stelt hen voor een enorme uitdaging.

Steeds meer informatie halen we digitaal naar ons toe. Mobiele en vaste netwerken hebben hierdoor te maken met een explosieve groei aan dataverkeer. Het einde hiervan is nog niet in zicht, voornamelijk door de toename van videodiensten die in steeds hogere resolutie worden gevraagd. Voor vaste netwerken betekent dit dat de standaard internetaansluiting via ADSL tot ongeveer 8 Mb per seconde niet meer volstaat. Aanbieders via de kabel (coax-kabel) zijn al vrij ver in het verglazen van hun netwerk, waardoor ze op dit moment al tenminste 100 Mb per seconde kunnen aanbieden, maar ook dit zal binnenkort niet meer voldoende zijn. Telecomoperators die een standaard telecommunicatie netwerk via zogenaamde twisted pair-koperkabel beheren zijn nog niet zo ver. Deels zijn ze bezig met het uitrollen van VDSL-gebaseerde netwerken, waarbij het glas tot aan de straatkast (fibre to the cabinet, FttCab) wordt gebracht en een snelheid tot ongeveer 100 Mb kan worden bereikt. Dit is nu misschien voldoende maar zeker niet toekomstvast. Daarnaast zijn ze in bepaalde gebieden glasverbindingen aan het aanbieden tot aan het huis, ook bekend als Fibre to the Home (FttH). Dit laatste gaat echter niet snel genoeg. In de afgelopen vijf jaar is ongeveer 25%[1] [138] van het land bereikt, waarbij waarschijnlijk bij het makkelijkste deel is begonnen. Een dergelijke aansluiting kost ook rond de 1000 euro per aansluiting wat een enorme investering betekent voor de operators. Een groot dilemma. Moeten ze nu snel een netwerk uitrollen waarvan ze nu al weten dat deze niet toekomstbestendig is, of, langzaam, een duur toekomstvast netwerk, waarbij misschien geen klant meer overblijft? In dit proefschrift wordt hier een planmatige oplossing voor geboden.

Elektriciteitnetwerken hebben met een ander dilemma te maken. Hier neemt niet alleen de vraag toe bijvoorbeeld door elektrische auto's, ook gaan de klanten (huishoudens) deels hun eigen elektriciteit opwekken. Zij worden ook wel *prosumers* genoemd, een combinatie van producer en consumer. Zonnepanelen worden gemeengoed, maar ook steeds meer elektriciteit-opwekkende cv-ketels (zogenaamde micro-WKK) en kleine windmolens worden bij huizen of lokaal in woonwijken geplaatst. Deze productie is ook nog eens zeer variabel, want afhankelijk van zon en wind. De toenemende vraag en de

---

[1] Stand september 2013.

onzekere lokale productie vraagt veel van de leveranciers. Hoe kunnen zij hun netwerk in balans houden? Oplossingen zoals zelf de opwekking flexibel maken, via (prijs-)prikkels de vraag van de consumenten beïnvloeden en smart grids introduceren worden als maatregelen onderzocht. Natuurlijk is er daarnaast nog de, zeer dure, oplossing van versterking en vernieuwing van de netwerkkabels. In dit proefschrift wordt gekeken naar een oplossing die ligt op tactisch niveau. Hoeveel van elk soort lokale opwekker moeten we waar plaatsen, zodat we het transportverlies en het verlies door vraag-aanbodmismatch minimaliseren?

Voor beide probleemgebieden wordt in dit proefschrift gekeken naar (tactische) planningvraagstukken, die allen zeer complex zijn om op te lossen. De operationele toepassing van de voorgestelde algoritmes bepalen mede de scope van de oplossingen die gepresenteerd worden in dit proefschrift. Door de wens vanuit de praktijk naar interactieve planningstools gaat de aandacht uit naar schaalbare methodieken die een voldoende goede oplossing genereren binnen een beperkte rekentijd.

Telecominfrastructuur

Bovenstaand dilemma voor telecomoperators vraagt om een uitrolstrategie die zoveel mogelijk aan de klantvraag voldoet, maar ook economisch op de lange termijn haalbaar is. Steeds, adhoc, aan de klantvraag voldoen kan te duur worden. Zo goedkoop mogelijk een toekomstvaste infrastructuur neerleggen kan ertoe leiden dat er op een gegeven moment geen klanten meer over zijn, aangezien ze allemaal zijn overgestapt naar de concurrent, 'de kabel', die eerder de gewenste internetsnelheid kan bieden. In dit proefschrift wordt een economisch model gepresenteerd dat allerlei strategien kan doorrekenen. Analyses met dit model leiden tot de conclusie dat een migratiepad van ADSL, via FttCab (VDSL-gebaseerd) en FttCurb (G.Fast-gebaseerd) naar FttH zowel qua investering voordelig is als de klanten vasthoudt.

Voor dit migratiepad is het van belang de opeenvolgende infrastructuren zoveel mogelijk in elkaars verlengde te bouwen, waarbij steeds bij de planning van de huidige oplossing rekening wordt gehouden met de karakteristieken en structuur van de volgende. In dit proefschrift wordt daarvan een voorbeeld gegeven. Vervolgens wordt voor elk van de migratiestappen (ADSL naar FttCab, FttCab naar FttCurb en FttCurb naar FttH) een planningsmethodiek gepresenteerd.

Voor de eerste migratiestap (stap 1), FttCab planning, wordt een planningsmethodiek gepresenteerd, in eerste instantie gebaseerd op drie stappen. Hier wordt eerst bepaald welke straatkasten voorzien moeten worden van actieve apparatuur voor de overgang van glasvezel naar koper. Doel hierbij is een minimum aantal kasten te voorzien, uit kostenoogpunt, waarbij wel een minimale dekkingsgraad van bijvoorbeeld 95% wordt gehaald. Dit houdt bij VDSL in dat 95% van de huizen binnen één kilometer over koperkabel vanaf de actieve apparatuur kunnen worden bereikt. Vervolgens moeten de geactiveerde straatkasten via een glasvezelring verbonden worden. Hiertoe worden eerst clusters gemaakt van een bepaalde maximale omvang, vanwege de capaciteit van de glasvezelring (stap 2) en wordt de route van deze ringen bepaald (stap 3). Alle drie de problemen zijn op zich al moeilijk oplosbaar (NP-hard). Voor elk van de drie problemen wordt daarom een heuristiek voorgesteld en wordt getoond hoe goed deze werkt. Van belang hierbij is dat we vooral geïnteresseerd zijn in een voldoende goede oplossing, die binnen enkele seconden te berekenen is. Voor stap 1 wordt daarom een greedy heuristiek ontwikkeld die voor alle testcases binnen één seconde een goede oplossing geeft. Voor stap 2 wordt een heuristiek ontwikkeld gebaseerd op Lloyd's algoritme voor het oplossen van zogenaamde k-means clusters. Voor stap 3 is het van belang glasvezelringen te creëren die elk niet tweemaal hetzelfde stukje traject gebruiken. Hiervoor wordt een heuristiek ontwikkeld die op basis van Dijkstra's algoritme en een TSP insertion algoritme snel een initiële oplossing vindt en dan op een greedy wijze de knelpunten oplost. Om een idee te krijgen hoe goed de driestaps aanpak werkt wordt vervolgens gekeken naar een methodiek die de tweede en derde stap simultaan oplost. Zoals te verwachten was geeft deze methodiek betere oplossingen, aangezien informatie over de routering wordt meegenomen in de clustering. Ook is zoals verwacht de rekentijd van deze methode langer, maar minder lang dan verwacht.

Voor de tweede migratiestap, FttCurb planning, worden de planningsopties geïnventariseerd. Hier worden de drie belangrijkste keuzes benoemd die spelen indien een FttCurb netwerk, gebaseerd op de technologie G.Fast, wordt aangelegd. Eerst moet bepaald worden of alle huizen worden aangesloten binnen de gewenste lengte, voor G.Fast ongeveer 200 meter, of slechts een bepaald percentage. Daarna moet bepaald worden of de nieuwe actieve nodes een capaciteitsbeperking hebben, of dat er ongelimiteerd kan worden uitgebreid op één lokatie. Tenslotte moet bepaald worden of de nieuwe actieve punten worden aangesloten met glasvezel via een ster- of boomstructuur, of via een ringstructuur. Als elke keuze twee opties heeft, dan heb zijn er totaal acht mogelijke configuraties van het netwerk. In dit proefschrift wordt voor elk van de acht opties de planningsmethodiek beschreven. Vervolgens wordt voor twee steden, Amsterdam en Den Haag, de planning via één van de bovenstaande methodes uitgevoerd waardoor een goede indicatie wordt gegeven van de kosten van dit netwerk. De resultaten van dit onderzoek kunnen ook gebruikt worden als validatie van het eerder genoemde (eenvoudigere) economische model.

De derde migratiestap tenslotte betreft de FttH planning. Hierbij wordt getoond hoe de lokatie van de nieuwe centrale node (de PoP) kan worden bepaald. Een eenvoudige heuristiek, de zogenaamde JMS-heuristiek, wordt op een aantal punten uitgebreid om meer details in deze methode mee te nemen. Vervolgens wordt gekeken hoe vanuit deze PoP de glasvezels naar de huizen moeten gaan lopen. Dit lijkt een triviale vraag, maar hier is veel winst te halen. Er wordt een nieuwe methode beschreven die bepaalt hoe de precieze loop van de glasvezels naar de woningen zo economisch mogelijk gekozen kan worden. Daarnaast wordt een uitbreiding op dit model gepresenteerd die de mogelijkheid biedt verder te kijken dan de aanlegkosten voor de aannemer of netwerkeigenaar. Het aanleggen van een dergelijk netwerk brengt namelijk veel overlast met zich mee. Straten worden tijdelijk afgesloten, winkels zijn onbereikbaar, groenvoorzieningen en bomen worden aangetast, enzovoort. Met de nieuwe methodiek kunnen deze kosten worden meegewogen en tevens kunnen mogelijkheden worden meegenomen om de vezels of buizen mee te leggen met andere infrastructurele werkzaamheden zoals herbestratingen en vernieuwingen van de riolering. Hierdoor komt er een gefaseerde aanleg tot stand die minder overlast oplevert.

Elektriciteitsinfrastructuur

De producenten van elektriciteit en de beheerders van de netwerken hebben zoals gezegd te maken met toenemende vraag en van een toename van sterk variabele aanbod door lokale opwekker zoals zonnecellen en windmolens. In dit proefschrift worden twee onderzoeken beschreven die kijken naar tactische planning van deze opwekkers. Hoeveel moeten er van welk soort geplaatst worden en waar moeten deze opwekkers geplaatst worden om verliezen te minimaliseren? Lokale opwekking kan transportverliezen verminderen omdat de elektriciteit niet ver getransporteerd hoeft te worden. Aan de andere kant zorgt deze opwekking voor grote variabiliteit in het aanbod, waarbij ook overschotten kunnen ontstaan die weer getransporteerd moeten worden, opgeslagen moeten worden met de bijkomende verliezen of die verloren gaan door gebrek aan vraag. Bij windmolens geldt vervolgens nog dat wanneer op een grotere schaal wordt gekeken, spreiding zorgt voor een lagere variabiliteit door een lagere correlatie tussen de wind op verschillende plaatsen. Als het waait in Middelburg, dan zal het ook waaien in Vlissingen. Dat het (even hard) waait in Groningen is een stuk minder zeker.

In het eerste onderzoek wordt gekeken naar één wijk en wordt bepaald welke combinatie van lokale opwekkers het minste verlies oplevert. Als elk huis kan kiezen uit meerdere soorten lokale opwekkers, welke keuze over al de huizen heen is dan, voor wat betreft transportverliezen, de beste keuze. Hiervoor wordt een Kwadratisch Mixed Integer Programmeringsprobleem gedefinieerd dat vereenvoudigd moet worden om het te kunnen oplossen. Afhankelijk van de verschillende aannames, zoals de vraag of het is toegestaan om overproductie naar andere districten te vervoeren, of het is toegestaan om een opslagsysteem te gebruiken en of er extra vraag is van elektrische voertuigen en warmtepompen, worden er verschillende resultaten uit het model verkregen. De verschillende resultaten geven wel aan dat toepassing van een optimale mix in de wijk energieverlies aanzienlijk kan verminderen. Een andere vraag is of het gebruik van de DG's, warmtepompen en elektrische voertuigen zorgt voor capaciteitsproblemen in het net. Als alle huizen beide soorten DG's hebben dan komen we geen problemen tegen van overbelasting. Bij het gebruik van warmtepompen en elektrische voertuigen is helaas problematischer. De meeste problemen met overbelasting komen van het gebruik van elektrische voertuigen, die zoveel van het netwerk eisen dat al zeer snel overbelasting van de kabels ontstaat. Het nadeel van het hebben van zowel micro-WKK systemen als zonnepanelen is dat ze elektriciteit genereren in dezelfde periode van de dag, ze zijn sterk gecorreleerd. Echter, elk type DG heeft een positieve eigenschap die kan worden gebruikt in specifieke situaties. Omdat zonnepanelen relatief kleine hoeveelheden elektriciteit produceren, kunnen ze gebruikt worden als aanvulling op andere opwekkers. En omdat micro-WKK veel elektriciteit genereren zijn zij vooral bruikbaar wanneer er een grote toename is in de vraag, bijvoorbeeld als gevolg van warmtepompen en elektrische voertuigen. Efficiënte opslagsystemen zullen natuurlijk erg helpen bij het verlagen van de verliezen door lokale generatie. Helaas zijn die nu nog niet van voldoende kwaliteit voorhanden. Deze resultaten tonen ook aan dat in plaats van willekeurig inzetten van DG's in de wijk, het beter kan zijn om de samenwerking tussen de huiseigenaren te bevorderen zodat grote reducties in energieverlies kunnen worden bereikt. Op deze manier zal het veel rendabeler zijn voor de hele wijk. Dit is politiek natuurlijk erg moeilijk en zal makkelijker kunnen gaat via woningbouwcorporaties.

In het tweede onderzoek wordt gezocht naar de optimale locaties om een bepaald aantal nieuwe windmolens te bouwen, om de verwachte transportverliezen te minimaliseren. Hier moeten twee afwegingen gemaakt worden: lokale of centrale plaatsing en spreiding of concentratie. Hiervoor is een wiskundig model geformuleerd, een zogenaamd Generation Expansion Planning probleem voor verliesminimalisatie. Dit is een tweestaps stochastische probleem. Dit probleem blijkt moeilijk oplosbaar. Daarom wordt een eenvoudige heuristiek gepresenteerd, samen met een procedure om windscenario's te simuleren voor steady state windenergie in meerdere gebieden. Hiermee worden resultaten voor het Nederlandse netwerk gegenereerd. Uit de resultaten kan geconcludeerd worden dat de afname van transmissieverliezen door het plaatsen van windmolens aanzienlijk zwakker is in het stochastische model dan voor het deterministische model. De stochastische windmolens worden verder uit elkaar geplaatst dan deterministische, maar interessante effecten doen zich slechts voor bij hogere penetratie windmolens. Voor de specifieke Nederlandse situatie geldt dat de optimale verdeling van windenergie over Nederland een hogere spreiding vertoont dan alleen verklaard kan worden uit de stochasticiteit. Het locatie-afhankelijke karakter van windenergie is zeer relevant voor het effect op de netwerkprestaties. Groningen en Friesland zijn de enige regio's waar de plaatsing van windturbines zou leiden tot vermindering van de transmissieverliezen in het HS-net. Maar bij de huidige omvang van windenergie lijkt een spreidingstrategie voor windturbines niet relevant, noch vanuit een commercieel oogpunt (variantie reductie van de totale productie voor de Nederlandse markt), noch vanuit een technisch oogpunt (het verminderen van de transmissie verliezen).

Curriculum Vitae

Frank Phillipson was born in 1973 in Purmerend. He studied Econometrics at the VU University Amsterdam, and wrote his Master's thesis in the field of Operations Research in 1996. This thesis was written at the R&D department of ASZ, part of GAK in Amsterdam and covered simulation of business processes to provide social security benefits. In the same year, he joined the Delft University of Technology to follow the Post-Doctoral program 'Mathematical Design Engineering' with a strong focus on application of Operations Research techniques in networks. The thesis of this program was written at Railned/Dutch Railroads on the subject of train traffic control in simulation of railtraffic.

From 1998 until 2002 he was employed at KPN Research. In 2003, KPN placed its research department in TNO, the largest applied research institute in the Netherlands, where Frank is currently working in the department 'Performance of Networks and Systems' as a senior scientist. There he is particularly working in the field of planning of ICT/telecom and electricity networks. In projects for customers the planning tools PlanXS, Tiger and Masterglass were built. In addition to this main topic, he has worked on projects for financial and economic models relating to telecom business.

www.ingramcontent.com/pod-product-compliance
Lightning Source LLC
Chambersburg PA
CBHW080242180526
45167CB00006B/2379